Automated Vehicles and MaaS

Automated Vehicles and MaaS

Removing the Barriers

Bob Williams
Senior Consultant
CSi (UK)
Oxfordshire
UK

Registered Offices
John Wiley & Sons, Inc., 111 River Street, Hoboken, NJ 07030, USA
John Wiley & Sons Ltd, The Atrium, Southern Gate, Chichester, West Sussex, PO19 8SQ, UK

Editorial Office
The Atrium, Southern Gate, Chichester, West Sussex, PO19 8SQ, UK

For details of our global editorial offices, customer services, and more information about Wiley products visit us at www.wiley.com.

Wiley also publishes its books in a variety of electronic formats and by print-on-demand. Some content that appears in standard print versions of this book may not be available in other formats.

Library of Congress Cataloging-in-Publication Data

Names: Williams, Bob, author. | John Wiley & Sons,
    Ltd., publisher.
Title: Automated vehicles and MaaS : removing the barriers / Bob Williams.
    Description: Hoboken, NJ : Wiley, 2021. | Includes bibliographical
    references and index.
Identifiers: LCCN 2020033769 (print) | LCCN 2020033770 (ebook) | ISBN
    9781119765349 (hardback) | ISBN 9781119765332 (adobe pdf) | ISBN
    9781119765387 (epub)
Subjects: LCSH: Intelligent transportation systems. | Automated vehicles. |
    Transportation and state.
Classification: LCC TE228.3 .W54 2021 (print) | LCC TE228.3 (ebook) | DDC
    629.28/30285–dc23
LC record available at https://lccn.loc.gov/2020033769
LC ebook record available at https://lccn.loc.gov/2020033770

Cover Design: Wiley
Cover Image: © Vladimir Kramin/Shutterstock

Set in 10/12pt WarnockPro by Straive, Chennai, India
Printed and bound by CPI Group (UK) Ltd, Croydon, CR0 4YY

C9781119765349_040521

I dedicate this book to my tolerant partner Isabelle, and my patient daughter Juliette, who have both enabled me the time in a limited amount of free time (for this is a very busy and exciting period for those of us who work on aspects of intelligent transport systems) to develop and write this book.

I dedicate this book also to those who provide a lot of time, skill and expertise, often largely unpaid, to develop standards that enable us to realise and benefit from the potential offered by intelligent transport systems, and who help to realise and achieve its safety of life potential.

Contents

Preface

There has been much hype, publicity and many optimistic claims regarding connected and automated vehicles and the concept of 'Mobility as a Service'. This book confronts, with a feet-on-the-floor approach, why these forecasts are so difficult to achieve, what is reasonably probable and achievable, and provides guidance, from an expert involved in the development of intelligent transport systems since 1988, on how to remove the barriers to the successful introduction of automated vehicles and Mobility as a Service.

Acknowledgements

I take this opportunity to acknowledge and thank my colleagues in CEN TC278 and its sister organisation ISO TC204 for their work in achieving practical, fair and open standards for intelligent transport systems. I thank them for their patience with me, and continual sharing of expertise that have helped me through a long career in this area, which has enabled me to pull together the information and experience necessary to write this book.

Table of Abbreviations

E & OE, abbreviations/acronyms in this book that are not detailed below may generally be understood to be the names of companies or trading entities. Abbreviations/acronyms used in this book are to be interpreted as follows:

2D	2 Dimension/2 Dimensional
3D	3 Dimension/3 Dimensional
3G	3rd Generation
3GPP	3rd Generation Partnership Project
4G	4th Generation
5G	5th Generation
AAM	Alliance of Automobile Manufacturers
ACC	adaptive cruise control
ACM	Association for Computing Machinery (USA)
ADAS	advanced driver assistance system
ADS	automated driving system
AEB(s)	automated emergency braking (system)
AEF	Association d'Economie Financière
AEV	automated and electric vehicles
AI	artificial intelligence
AMC	American Motor Company (American Motors) (USA).
APEC	Asia-Pacific Economic Cooperation (Organisation)
API	application program interface
ARK-Invest	(Company name
ARC-IT	Architecture Reference for Cooperative and Intelligent Transport
ASEAN	association of south east asian nations
ASN.1	abstract syntax notation 1
Auto ISAC	automotive information sharing and analysis centre
AV	automated vehicle
AVO	AV operator
BAST	Bundesanstalt für Straßenwesen (German: Federal Highway Research Institute)

BMW	Bayerische Motoren Werke GmbH
BSM	basic safety message
CA	certificate authority
CALM	communications architecture for land mobiles
CAM	cooperative awareness message
C-ART	Name of JRC project (Coordinated Automated Road Transport)
CAMP	Crash Avoidance Metrics Partnership
CAV	connected and automated vehicle(s)
CCAM	cooperative connected automated mobility
CCMS	C-ITS credential management system (EU)
CES	Consumer Electronics Show (Consumer Technology Association trade show)
CEO	chief executive officer
C-ITS	cooperative-ITS
CMS	credential management system
CNECT	Name of European Commission Directorate General (DG) for Communications Networks, Content and Technology.
CONVERGE	Communication Network Vehicle Road Global Extension (EC Project)
C-ROADS	connected-roads (EC DG MOVE Initiative)
CRL	certificate revocation list
CVT	continuously variable transmission
C=V2X	communications – vehicle to anything
DARPA	Defense Advanced Research Projects Agency (USA)
DAS	driving automation system (sae) or driving assistance system (others)
DATEX	data exchange standard for exchanging traffic information between traffic management centres, traffic service providers, traffic operators and media partners
DDT	dynamic driving task
DENM	decentralized environmental notification message
DfT	Department for Transport (UK)
DoT	Department of Transport (USA and others)
DG	Directorate General (European Commission)
DG-CNECT	DG- Communications Networks, Content and Technology (EC)
DG-GROW	DG- for Internal Market, Industry, Entrepreneurship and SMEs (EC)
DG-INFSO	DG- Information society and Media (EC: Now DG-CNECT)
DG-MOVE	DG- Mobility and Transport (EC)
DSRC	dedicated short range communication
DSS	driver support system
DSSAD	data storage system for automated driving vehicles
EC	European Commission
ECTL	European certificate trust list
ECU	electronic communications unit

EDR	event data recorder
EEA	European Economic Area
EFTA	European Free Trade Agreement
ENISA	European Union Agency for Networks and Communication
ESC	electronic stability control
ETSC	European Transport Safety Council
ETSI	European Telecommunications Standards Institute
ETRAC	European Road Transport Research Advisory Council
EU	European Union
ExVe	extended vehicle
EZ-GO	(Trade name [Renault])
FHWA	Federal Highway Administration (USA)
FRAV	functional requirements for automated vehicles
FSD	Full Self-Driving (Tesla product name)
GAD	(UK) Government Actuaries Dept
GDP	gross domestic product
GDPR	General Data Protection Regulation (EU)
GEAR 2030	name of high-level group on the competitiveness and sustainable growth of the automotive industry in the European Union (EC)
GLOSA	green light optimal speed advisory
GM	General Motors
GMC	governance management committee
GNSS	global navigation satellite system
GPS	global positioning system (US Department of Defence GNSS)
GRB	UNECE WP 29 former name of working party subcommittee on noise and tyres
GRBP	UNECE WP29 Working Party on Noise and Tyres (Groupe Rapporteur Bruit et Pneumatiques)
GRE	UNECE WP 29 Working Party on Lighting and Light-Signalling (Groupe Rapporteur Electrique)
GRPE	UNECE WP 29 Working Party on Pollution and Energy (Groupe Rapporteur pollution et énergie)
GRRF,	UNECE WP 29 Working Party on Brakes and Running Gear (Groupe Rapporteur Freins & roulants)
GRSG	UNECE WP 29 Working Party on General Safety Provisions (Groupe Rapporteur Securite Generale)
GRSP	UNECE WP 29 Working Party on Passive Safety (Groupe Rapporteur sécurité passive)
GRVA,	UNECE WP 29 Working Party of automated vehicles (Groupe de Rapporteurs pour les Véhicules Autonomes)
HARTS	Harmonised Architecture for Transport Systems (HTG initiative)
HMI	human >< machine interface
HOV	high occupancy vehicle
HP	Hewlett Packard

HTG	Harmonization Task Group (a collaboration of US DOT, the EU Joint Research Council (JRC), EC DG CNECT and the Transport Certification Authority (TCA) of Australia),
ICT	Information and Communications Technologies
IEEE	Institute of Electrical and Electronics Engineers
IGEAD	Informal Group of Experts on Automated Driving (UNECE WP29)
IMOVE	(name of EC project)
IMS	IP multimedia subsystem
INSPIRE	(Name of EU Directive: Directive 2007/2/EC of the European Parliament and of the Council of 14 March 2007 establishing an Infrastructure for Spatial Information in the European Community
IP	internet protocol
IP	intellectual property
IPR	intellectual property rights
ISA	intelligent speed assistance
ISO	International Standards Organisation
ITF	International Trade Forum
ITS	intelligent transport systems
ITS-G5	intelligent transport systems – 5 ghz wifi for transport communications
ITS-S	intelligent transport systems – station
ITS-SU	intelligent transport systems – station unit
JRC	Joint Research Centre (of European Commission)
LA	local authority
LAN	local area network
LKA	lane keeping assistance
LOM	la loi d'orientation des mobilités or, mobility orientation law) (France)
LTE	Long Term Evolution (3GPP)
MaaS	Mobility as a Service
METR	Management of Electronic Traffic Regulations
MOAT	Managed Optimisation Architecture for Transportation
NASA	National Aeronautics and Space Administration (USA)
NHTSA	National Highway Traffic Safety Administration (USA)
NIS(D)	Network and Information Security Directive (EC)
NTSB	National Transportation Safety Board (USA)
OD	operational design or operational domain
ODD	operational design domain
OECD	Organisation for Economic Co-operation and Development
OEDR	object and event detection and response
OEM,	original equipment manufacturer
OICA,	Organisation Internationale des Constructeurs d'Automobiles
OTA	over the air
PHV	private hire vehicle(s)
PKC	public key certificate

PKI	public key infrastructure
PMC	policy management committee
PS-LTE,	packet switched-long-term evolution (3GPP)
PT	public transport/public transit
PTA	public transport authority(ies)/ public transit authority(ies)
PTO	public transport operator
PTW	powered two-wheeler
RA	registration authority
RAN,	radio access network
RB	resource block
RCA	root certificate authority
RER	Réseau Express Régional (Paris)
RGMC	regional governance management committee
RO	road operator
RTMS	road traffic management system
R&D	research & development
SAE	Society of Automotive Engineers (USA)
SAEV	shared autonomous electric vehicles
SCMS	security credential management system
SDO	standards development organization
SMMT	Society of Motor Manufacturers and Traders (UK)
SSP	system security plan
SUV	sports utility vehicle
TBD	to be decided
TC	technical committee
TCA	Transport Certification Australia
TfL	Transport for London
TIP	travel information provider
TLM	trust list manager
TMC	traffic management centre
TNC	transportation network company
TN-ITS	Traffic Navigation-ITS (Organisation, EU)
TOS	travel optimisation service
TRAMAN21	Traffic Management for the 21st Century (EC Project name)
TSP	travel service provider
UI	user interface
UK	United Kingdom of Great Britain and Northern Ireland
UMTS	Universal Mobile Telecommunications Service
UNECE	United Nations, Economic Commission for Europe
USA	United States of America
UU	user equipment to the UMTS terrestrial radio (interface)

UVAR	urban vehicle access regulations/restrictions
V2I	vehicle to/from infrastructure
V2V	vehicle to/from vehicle
V2X	vehicle to/from anything
VACS	vehicle automation and communication systems (TRAMAN21 term)
VANET	vehicular ad hoc networks
VDA	Verband Der Automobilindustrie E.V. (German Automobile Industry Association)
VKT	vehicle kilometres travelled
VMS	variable message sign
VRU	vulnerable road user
VW	Volkswagen
WAVE	wireless access in vehicular environments
WG	working group
WHO	World Health Organisation
WLAN	wireless local area network
WP	working party
WWII	World War II

1

The Promise and Hype Regarding Automated Driving and MaaS

Figure 1.1 1950s image of highway of the future: *Popular Mechanics* magazine. Source: Popular Mechanics magazine.

1.1 The Promise

It is possible that the fully automated car was first seen in a road safety awareness film 'The Safest Place' (1935). *'The vehicle always stays in its lane, never forgets to signal when turning, obeys all stop signs and never overtakes on dangerous corners'* Kröger (2016).

By 1939, at the World's Fair, General Motors 'Futurama' featured a model of future transport systems with automated highways in an imagined world of 1960 Weber (2014). Please note with a smile, futurologists usually overestimate the speed of development and uptake of their subject (Figure 1.1).

Advances in computer technology have seen the rapid development of automation over the past 50 years. Combined with innovative engineering, this has led to developments from unmanned aerial vehicles (UAVs/drones) to armed robotic rovers. The US Armed Forces and DARPA built on the philosophy of 'development through competition' based on the early twentieth-century Orteig Prize (US$25 000) offered in 1919 by French hotelier Raymond Orteig for the first nonstop flight between New York City and Paris that helped prod the development of air flight, and that spurred Charles Lindbergh

Automated Vehicles and MaaS: Removing the Barriers, First Edition. Bob Williams.
© 2021 John Wiley & Sons Ltd. Published 2021 by John Wiley & Sons Ltd.

to make his solo flight across the Atlantic Ocean in 1927. DARPA have sponsored a number of competitions to accelerate the development of everything from automatic weaponry to private sector space flight.

In 2004, DARPA established the 'Grand Challenge', a competition designed to encourage the development of technologies needed to create the first fully autonomous ground vehicles.

The first Grand Challenge took place on 13 March 2004 and involved 15 self-driving ground vehicles navigating a 228 km (142 mi) course across the desert in Primm, Nevada (https://www.wired.com/story/autonomous-car-chaos-2004-darpa-grand-challenge/). The prize was $1 million but the desert course proved to be too hard. No team finished the course, and the prize went unclaimed.

The second event was held on 8 October 2005 in southern Nevada with 5 of the original 195 teams completing the 212 km (132 mi) and the $2 million prize was won by Stanford University.

For the third event, held in November 2007, DARPA extended the challenge to include a mock urban environment. Driving in traffic and typical vehicle manoeuvres and highway crossings were involved. Tartan Racing, a team from Carnegie Mellon University in Pittsburgh, Pennsylvania, claimed the $2 million prize with their vehicle 'Boss', a converted Chevrolet Tahoe.

Thus the race to the development of automated vehicles kicked off and was incentivised, and its progress has only accelerated thereafter.

* * * * *

We already live in a world where vehicles are to some extent 'connected'. New model vehicles in Europe have a system called 'eCall', which automatically contacts and puts the occupants of the vehicle in touch with the emergency services in the event of an accident. Volvo Assistance, BMW Connected Drive, GM Onstar, Mercedes 'Me' and 'Rescue' as well as Citroen Assistance are examples of breakdown, emergency and driver support systems that are connected to resources outside of the vehicle, connected by 2G/3G/4G, and soon to be 5G, mobile telephony.

The modern vehicle also 'connects' to its environment in many ways, largely through sensors, to assist with the driving experience. Electronic stability control (ESC) is now mandatory on all new cars sold in Europe. Lane-keeping systems (LKS), adaptive cruise control (ACC), automated emergency braking (AEB), and intelligent speed assistance (ISA) systems are increasingly commonplace, as are automatic headlight dipping, traction control, tyre-pressure monitoring, etc. It is thought-provoking to consider that most of what these systems do is to use technology to compensate, to some extent, for human error, often taking some control away from the driver under certain circumstances.

Modern sat-nav systems download and take into account dynamic congestion and traffic incident information in their route planning, and guidance by sat-nav providers communicate this data wirelessly to the on-board sat-nav system. Researchers and developers are close to the fruition of car-to-car and car-to-infrastructure communication developments, that will enable a truly 'connected' vehicle ('cooperative ITS' or 'C-ITS' as it is known in the trade).

Moving beyond such connectivity-enabled functions, attention has now moved to the often misnomered 'autonomous' vehicle that will understand its environment and the requirements of its passengers, and the requirements of the road infrastructure, and operate the vehicle without the assistance of a driver (more correctly called the 'automated' vehicle). It will also 'learn' to react and adapt to different situations during the entire driving process.

Over the next 10–50 years, the transport sector may expect to undergo a significant change, and potentially, transformation, as connected and automated vehicle technology is introduced.

With the impending take-up and spread of cooperative ITS (C-ITS) systems in vehicles, informative features will be complemented by, or evolve, cooperative features that will enable vehicles to interact with each other and with the surrounding infrastructure (i.e. vehicle-to-vehicle V2V and vehicle-to-infrastructure V2I communication). Full-scale deployment of C-ITS enabled vehicles that communicate with other vehicles concerning potentially dangerous situations and communicate with local road infrastructure is expected in the near term, and indeed may be required by regulation (for new vehicles), at least in Europe, by the early 2020s.

Many future projections estimate that by 2025, high automation driving will be available on highways and by 2030 in cities. The EC's Joint Research Centre further forecasts the year 2050 as a realistic timescale for the transition to a future mobility paradigm.

In order to summarise the potential of automated driving, ETSC, the European Transport Safety Council refers to the European Road Transport Research Advisory Council, who have summarised "safety and the potential to reduce accidents caused by human error" is one of the main drivers for higher levels of automated driving. "Automated driving can therefore be considered as a key aspect to support several EU transport policy objectives including road safety".

Automated and connected vehicles have the easy to understand potential to substantially reduce road accidents, traffic congestion, traffic pollution and energy use, and are therefore seem attractive to and are often encouraged/incentivised by governments. Automated vehicles also promise to increase productivity and comfort and to facilitate a greater inclusion in the mobility of specific groups of individuals such as disabled or elderly. But other projections for instantiation in other paradigms predict the opposite in respect of automated vehicles, i.e. an increase in traffic congestion, an increase in traffic pollution and an increase in energy use, and other studies indicate that, particularly in the early years, may also actually increase accidents (even though the accidents may not be the fault of the automated vehicle, but how others react to it).

What is clear is that we are not dealing solely with the efficiency of vehicle control functions to automatically drive vehicles, but with the road transport system as a whole, which is a complex one where road users, vehicles and infrastructure interact with each other and millions of decentralised decisions are taken every second, by human drivers, and other road users, and within which automated vehicles will have to operate as a managed part of the system.

Automated and connected vehicles potential contribution to reduce road accidents is achieved primarily by eliminating human errors, which are a contributing factor in a vast majority of road accidents. And it is also generally recognised that most accidents occur due to risks that human drivers continuously take (consciously and unconsciously) as a result of collective experience gained in more than 100 years of driving activities, and of past driving experiences of the driver over his/her lifetime.

But, as the C-ART report (EC JRC 2017) points out, *'if on the one hand these risks generate road accidents with all their negative consequences, on the other hand these risks usually have a positive effect on the capacity of the road transport system.*

The introduction of automated vehicles, which by definition will be designed to minimize the risk of accidents, could therefore have a negative effect on road capacity especially in a transition period where a mix of conventional and automated vehicles will be sharing the same infrastructure'.

Features most prominently making progress at the moment are so called ADAS (advanced driver assistance system) functions. These systems generally refer to systems such as automatic braking, collision protection and emergency assistance. As the technologies evolve and mature, ADAS will soon evolve into part of the automated driving package.

Telematics and infotainment services that are already in place in modern vehicles use connectivity features. These services will expand, probably rapidly, sometimes as part of the selling options, sometimes as subscription services, and sometimes simply through smartphone apps integration.

1.2   What Do We Mean by the Term 'Automated Driving'?

Automated driving combines a wide range of technologies and infrastructures, and importantly, connectivity. Automated driving should also be seen within the broader context, not just of taking the driving function away from the user of the vehicle, but also enabling new disruptive paradigms for mobility that may change the way we travel, change the vehicle ownership paradigm, change where we choose to live, etc., especially in urban environment.

Automated vehicles are those which blend autonomous control with communication with other vehicles and with the infrastructure in order to control and manage vehicle movements from start point to destination without direct driver input. Automated vehicles use a mix of on-board sensors, cameras, Global Navigation Satellite Systems (GNSS), and telecommunications to obtain information in order to make their own judgements regarding safety-critical situations and the general management of the journey.

Vehicle manufacturers frequently use the term 'autonomous vehicle' (the definition of autonomous is 'having the freedom to govern itself or control its own affairs' or similar), although what they go on to describe is clearly a vehicle communicating with its environment, and not one solely relying on the vehicle's own systems and without communicating with other vehicles or the infrastructure. The author observes that this highlights the shortcomings of the vehicle manufacturer's vehicle centric visions of the paradigm for automated driving, which is focussed on the vehicle controlling its movements through the environment, rather than the reality that the vehicle is only allowed to operate within the limits set by traffic management control and regulations.

While the auto manufacturers largely see the vehicle controlling its movements as the controller of the paradigm, in instantiation, automation has several key stakeholders, starting with the road authority or local/city administration (because they provide the road infrastructure, regulations, street equipment and signage and in the future may operate transport optimisation services). Transport optimisation services are key stake-holders (because they dynamically control all traffic movements through the network). Of course, the automotive manufacturers, and the users of the vehicle, while not being the sole controllers of the automated vehicle, are key stakeholders. Similarly, as these vehicles are communicating and receiving communications, communications providers are also key stakeholders, likewise, transport managers (road, rail, metro, parking facility operators, bus, cycle and scooter share, etc.).

With many 'Mobility as a Service' (MaaS) paradigms involving automated driving, vehicle-sharing service providers are another potentially key stakeholder. There are also other stakeholders such as technology providers, insurance companies, and aftermarket service providers. And there will be other actors involved such as driver clubs/associations, universities and research centres.

In Europe, EU Regulation adopts UNECE (United Nations, Economic Commission for Europe) type approval regulations to provide (and require) access to basic raw (on board diagnostic) data from the vehicle by regulation, and there is serious consideration as to what additional data should – will have to – be made available for cooperative safety services, and to enable a fair and open after-market (although the vehicle manufacturers continue to try to control access to data, and where possible use it to generate an income stream). In North America the situation is currently less regulated, therefore more unclear, but one way or another these 'connected' services will continue to develop, and push towards the automation of more and more services.

1.3 The Hype

Most of the leading vehicle manufacturers are bullish about the prospects for self-driving vehicles. So rather than my version of what they are claiming, I simply turn to what is on their websites:

The Ford Motor company website (2019) stated:

Looking Further

Ford will have a fully autonomous vehicle in operation by 2021

No driver required. Thanks to Ford, that statement will be possible in 2021, the year that we will have a fully autonomous vehicle in commercial operation. To make this possible, we have partnered or invested with four different technology companies, along with doubling our Silicon Valley presence.

The effort to build fully autonomous vehicles by 2021 is a main pillar of Ford Smart Mobility: our plan to be a leader in autonomy, connectivity, mobility, customer experience, and analytics. The vehicle will operate without a steering wheel, gas pedal or brake pedal within geo-fenced areas as part of a ride sharing or ride hailing experience. By doing this, the vehicle will be classified as a SAE Level 4 capable-vehicle, or one of High Automation that can complete all aspects of driving without a human driver to intervene.

The SAE International six levels of automation rating system is used by the U.S. Department of Transportation to classify a vehicle's automation capabilities. The system starts at Level 0 – No Automation – which is defined as a vehicle that requires a human driver for all aspects of the driving task, and goes up to Level 5 – Full Automation – in which a vehicle can perform all driving tasks, no matter the environmental or roadway conditions. By mass producing a Level 4 capable vehicle, Ford will have achieved the highest level of automation by any automotive maker to date.

In order to reach this ambitious goal, Ford has committed to expanding its research in advanced algorithms, 3-D mapping, radar technology and camera sensors. To help accelerate the development of these new technologies, we have announced four key investments and collaborations with Velodyne, SAIPS, Nirenberg Neuroscience LLC and Civil Maps. These companies bring their own unique skill sets and experiences to the table, and have proven to be dedicated to making the world a better place through their technological endeavors.

Since becoming the first automaker to begin testing fully autonomous vehicles inside Mcity, the University of Michigan's simulate urban environment, Ford has made enormous strides in researching how these vehicles operate in hazardous conditions, such as snow and complete darkness. Over the next two years, we will have tripled our autonomous vehicle test fleet to 30 Fusion Hybrid sedans in 2017 and will have 90 by 2018. These sedans will be taking the roads in California, Arizona, and Michigan for extensive development and testing.

In addition to the extensive testing of these vehicles and intensive collaboration with outside partners, Ford is focusing on expanding its Silicon Valley presence by creating a dedicated campus in Palo Alto to ensure that these innovations will be made. The Ford Research and Innovation Center that was initially created in 2015 will have two new buildings and 150,000 square feet of work and lab space added, and the current Palo Alto staff of 130 people will be doubled by the end of 2017. (Source: Looking Further, Ford will have a fully autonomous vehicle in operation by 2021, Autonomous 2021. © 2019, Ford Motor Company.)

The Daimler/Mercedes (Figure 1.2) websites (2019) states :

Prototypes such as the Mercedes-Benz S-Class S 500 INTELLIGENT DRIVE, the F 015 Luxury in Motion or the Future Truck 2025 show that the technical conditions for autonomous driving are already well established. And demonstrations of our intelligent Highway Pilot in the Freightliner Inspiration Truck in Nevada (USA) and a series production Mercedes-Benz Actros in Germany have proven that it is ready for autonomous driving on public roads.

The required sensors and cameras have long been used in series production vehicles and undertake increasing numbers of tasks on the driver's behalf. Today's discussion no longer revolves around whether the technology will deliver on its promise but whether people

Figure 1.2 Mercedes F 015 concept AV at Detroit Motor Show (2015). Source: Mercedes-Benz.

want what the technology can deliver. And whether society and legislators are ready for this "revolution in automobility."

When it comes to passenger transport, autonomous driving ensures more safety, more comfort and more mobility. A purpose-designed research vehicle, the F 015 Luxury in Motion, shows what an autonomous Mercedes-Benz may look like in the future. We are convinced that the car can be more than just a means of transport: we see it as a private retreat that offers more freedom. Because autonomous driving allows us to use our time on the move as we wish. Naturally the design of this private retreat reflects the Mercedes-Benz brand.

When it comes to freight transport autonomous driving brings additional advantages. More efficiency: a more steady flow of traffic reduces fuel consumption and emissions. Better coordination of all processes: thanks to the connection to telematics solutions for fleets, routes and trips, diagnosis and servicing can be better planned. And what difference will it make for the drivers? On the one hand, time pressure will be reduced. After all, if all partners are informed about the progress of the journey in real time, there is no reason to explain a delay. On the other hand, drivers can "hand over control" on monotonous stretches of road. Anyone picturing a driver dozing in a moving truck at this point is very much on the wrong track, however. The driver is a key part of the system. In certain traffic situations, for example on motorways and rural roads, in city traffic and when connecting semitrailers and making deliveries, the driver must retain control of the truck.

Autonomous mobility will bring about major changes and involve many different parties. Psychological barriers must be overcome in the same way that social acceptance must be gained. This requires a sound legal basis across country borders that regulates autonomous traffic and covers questions of liability in the event of a collision.

French manufacturer Renault, says on its website (2019):

"Our goal is to provide our customers with models that feature a delegated autonomous driving mode from 2020 onward. This technology will make driving safer and more pleasant while also freeing up time for drivers." Laurent Taupin, Chief Engineer – Autonomous Vehicle

From ADAS to Autonomous Driving

Groupe Renault currently offers advanced driver assistance systems on its vehicles. These ADAS improve safety and act for the most part without human input, as is the case for automatic emergency braking (AEBS). They serve as a gateway to autonomous vehicles, even though they are initially only there to provide assistance to the driver, who remains in charge of the vehicle.

Eyes off/hands off technology

"Eyes Off/Hands Off" technology is a form of autonomous driving with no driver supervision. When drivers delegate driving activities, they no longer need to watch the road or have their hands on the steering Wheel - driving is now fully delegated to the vehicle. This feature is intended for the most boring kinds of driving - driving in stop-and-go traffic, for instance - and only on approved highways.

Eyes off/hands off mode

When autonomous driving mode is activated, a set of sensors monitor the road and provide 360° surveillance of the vehicle: lidars (long-range laser scanners), long-range frontal radar, medium-range corner radar, frontal digital cameras, four 180° digital cameras, an ultrasound belt and more. The data collected by these sensors is analyzed by the many embedded software "brains" that tell the vehicle what to do.

Figure 1.3 Screenclip from Renault website. Source: Renault.

Eyes off/ hands off benefits

With the autonomous "Eyes Off/Hands Off" mode, Renault's goal is to change the experience of riding in cars, making it more pleasant, more interesting and safer. This will significantly reduce the risk of accidents! Trips will be less stressful and more productive. Drivers can make better use of their time by using in-vehicle connectivity to answer emails or watch videos. They can do so safely, outside conditions permitting, as long as applicable laws and regulations are followed, once these are updated to authorize these new features. (Source: Renault, Autonomous Vehicle, © 2019, Renault.)

EZ-GO Concept

The robot-vehicle that reinvents the relationship between space and time. In an urban environment, EZ-GO Concept is the first incarnation of autonomous, connected and shared mobility using an electric engine, without a steering wheel or a driver (Figure 1.3).

Mobility on demand for everyone

EZ-GO Concept is a "robot-vehicle": both a vehicle and a transport service but also plays an integral part of the urban ecosystem. It has a positive impact on city life by providing mobility that is more respectful of the environment. The use of this autonomous vehicle for public transport takes it to the forefront of a new urban way of life. A facilitator for everyday life, EZ-GO Concept offers a genuinely connected and customised experience to its passengers. With frontal doors, limited speed and autonomous driving, EZ-GO Concept puts the safety of passengers at the forefront. The AD lighting signature, messages from the illuminated scrolling displays and the vehicle's exterior sounds ensure the safety of pedestrians. (Source: Renault, EZ-GO Concept, © 2019, Renault.)

The Verge, an online multimedia publication designed to examine how technology will change life in the future, recently posted (2019)

General Motors plans to mass-produce self-driving cars that lack traditional controls like steering wheels and pedals by 2019, the company announced today. It's a bold declaration for the future of driving from one of the country's Big Three automakers, and one that is sure to shake things up for the industry as the annual Detroit Auto Show kicks off next week.

The car will be the fourth generation of its driverless, all-electric Chevy Bolts, which are currently being tested on public roads in San Francisco and Phoenix. And when they roll off the assembly line of GM's manufacturing plant in Orion, Michigan, they'll be deployed as ride-hailing vehicles in a number of cities.

"It's a pretty exciting moment in the history of the path to wide scale [autonomous vehicle] deployment and having the first production car with no driver controls," GM President Dan Ammann told The Verge. "And it's an interesting thing to share with everybody."

"THE FIRST PRODUCTION CAR WITH NO DRIVER CONTROLS"

The announcement coincides with the tail end of CES, where a number of big companies announced their own plans to deploy autonomous vehicles, and right before the Detroit Auto Show, where the industry will have on display all the trucks and SUVs that make its profits.

By committing to rolling out fully driverless cars in a shortened timeframe, GM is seeking to outmaneuver rivals both old and new in the increasingly hyper competitive race to build and deploy robot cars. Ford has said it will build a steering-wheel-and-pedal-less autonomous car by 2021, while Waymo, the self-driving unit of Google parent Alphabet, is preparing to launch its first commercial ride-hailing service in Phoenix featuring fully driverless minivans (though still with traditional controls).

Unlike those other companies, GM provided a sneak peek at how its new, futuristic cars will look on the inside. In some ways, [it's] the vehicular version of a [Rorschach] inkblot test. The bilateral symmetry of the interior looks both unnerving and yet completely normal at the same time. Instead of a steering wheel, in its place is blank real estate. Under the dash, more empty space.

The automaker submitted a petition to the National Highway Traffic Safety Administration for permission to deploy a car that doesn't comply with all federal safety standards. Ammann said the company wasn't seeking an exemption from the Federal Motor Vehicle Safety Standards — something the government caps at 2,500 — just a new way around a few of the requirements.

GM is proposing to "meet that standard in a different kind of way," Ammann said. "A car without a steering wheel can't have a steering wheel airbag," he said. "What we can do is put the equivalent of the passenger side airbag on that side as well. So its to meet the standards but meet them in a way that's different than what's exactly prescribed, and that's what the petition seeks to get approval for." (Source: Andrew J. Hawkins, GM will make an autonomous car without steering wheel or pedals by 2019, Jan 12, © 2018, Vox Media.)

The Volvo website (2019) offers:

What is autonomous driving?

We believe that mobility should be safer, sustainable and more convenient. For Volvo Cars, technology should make people's lives easier. That's why our approach to autonomous driving is all about the people that will use them. Our future cars will be able to navigate without human input, equipped with sensors that read the surroundings, adapting to changing traffic conditions.

Unsupervised driving

In unsupervised autonomous mode, a vehicle performs all the driving because it is safe to rely on the technology to steer, brake and accelerate. People on board the vehicle are not expected to have control of the car.

Why autonomous cars?

Unsupervised autonomous cars will revolutionise society, boost global economies and transform the way we manage our time. As the biggest change to personal mobility since the invention of the car 130 years ago, we think there's a lot to look forward to. At Volvo Cars we believe that our first unsupervised autonomous vehicles will be in the market by 2021. What makes our approach to autonomous driving so unique is that we focus on people – not just on technology (Source: Volvo, Autonomous Driving. © 2019, Volvo Car Corporation.)

Even the austerity-following UK Chancellor of the Exchequer, Phillip Hammond, not one noted for his optimism, quoted early in 2019, that…. "the autonomous car, probably powered by an electric motor, will be on British roads, unsupervised, by 2021." (*The Guardian* 2019)

Anthony Cuthbertson writing for the middle of the road UK newspaper, *The Independent*, shortly after, (2019) reported:

Driverless cars to be rolled out on UK roads by end of 2019, government announces

'Key priority must be ensuring cyber security defences are deployed so this fantastic, ground-breaking technology does not fall victim to hackers,' …….

Self-driving cars without a human supervisor will be tested on public roads in the UK by the end of the year, under government plans.

Fully driverless trials have previously only taken place on a limited scale in the US and Europe.

The Department of Transport said the move towards advanced trials would push the UK to the forefront of the industry.

'Thanks to the UK's world class research base, this country is in the vanguard of the development of new transport technologies, including automation,' said Jesse Norman, the transport minister.

'The government is supporting the safe, transparent trialling of this pioneering technology, which could transform the way we travel.'

'The UK has a rich heritage in automotive development and manufacturing, with automated and electric vehicles set to transform the way we all live our lives,' said Richard Harrington, the automotive minister.

Uber plans self-driving bicycles and scooters.

'We need to ensure we take the public with us as we move towards having self-driving cars on our roads by 2021. The update to the code of practice will provide clearer guidance to those looking to carry out trials on public roads.'

Advanced driverless trials on UK roads by the end of 2019 will also help the government keep to its commitment of having self-driving vehicles on UK roads by 2021, ministers said. (Source: Anthony Cuthbertson, Driverless cars to be rolled out on UK roads by end of 2019, government announces, 6 February. © 2019, The Independent.)

And even the normally staid departments of government are getting excited. The UK Government Actuaries Dept (GAD), in September 2017 published:

Self-Driving Cars

Once just viewed as part of science fiction, self-driving cars, perhaps more correctly referred to as connected and autonomous vehicles (CAVs), are already here in various forms. Connected vehicles are those which are able to communicate with their surroundings providing information on road, traffic and weather conditions. The next level is automation where the vehicle uses its connection to assist the driver, examples include autonomous emergency braking, adaptive cruise control and park assist. Testing is now also well under way for vehicles that take full control from start to finish – fully autonomous vehicles.

The vast majority of road accidents relate to human error and reducing such accidents is projected to contribute £2 billion of savings to the economy by 2030. The total projected economic benefits from all sources are in excess of £51 billion. These figures highlight the importance of developing this technology.

The UK Society of Motor Manufacturers and Traders (SMMT) issued a report (2019). This report 'offers a detailed assessment of connected and autonomous vehicle (CAV) development, and crucially deployment, in the UK'. The report envisages the economic benefit to the UK from the deployment of CAVs (connected and automated vehicles), to be in the region of £62 billion per annum by 2030. The SMMT report forecasts the benefits, just for UK, to be (Figure 1.4):

'The emergence of CAVs will bring unprecedented change to the automotive industry worldwide. More than 18 million new automated vehicles are expected to be added to the global motor parc by 2030, significantly changing the way people commute. Over the next decade, for instance, new mobility modes such as automated shuttles could address gaps in first and last mile mobility'.

................'As this happens, disruption is likely to occur across traditional, ownership focused vehicles as well as shared mobility services such as taxis and shuttles. For example, it is estimated that there will be a 15% reduction in all collisions across major markets, including in

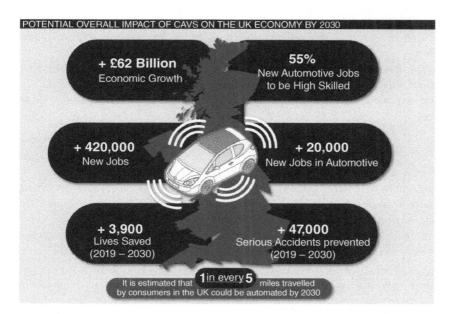

Figure 1.4 SMMT Potential overall impact of CAV's on the UK economy (screenclip from public website). Source: The UK Society of Motor Manufacturers and Traders (SMMT) issued a Report "Connected and Autonomous Vehicles 2019 Report / Winning the Global Race to Market. © 2019, SMMT".

Europe and North America, within a span of 10 years of Automated Emergency Braking (AEB) being mandated in Europe (expected between 2021 and 2025).'

The report assesses the Global situation and reports that 'milestones reached in nations around the globe in 2018'.

GLOBAL OVERVIEW OF CAV DEVELOPMENT

Several countries are currently exploring the impact of AVs on their cities and highways. Crucial to building this understanding is real world testing. Accordingly, many test beds have been set up to help collect vital data on how AVs interact with their surroundings, with a series of major milestones reached in nations around the globe in 2018.

Europe

- *AV testing on public roads was legalised in several countries, including France, Germany, the Netherlands, Norway, Sweden and the UK.*
- *A legal framework for AVs, including driving licence equivalents for self-driving vehicles, was reviewed*
- *in the UK, the Netherlands, and Germany. The UK Parliament approved further clarity on liability in self-driving mode, introducing the world's first insurance legislation for AVs (2019)*
- *Oxbotica, a UK start-up focused on AV software stack and related services, began on-road testing for Level 4 automated grocery delivery vehicles, taxis and shuttles at locations across the UK.*

North America

- *California passes state level approval for driverless vehicle testing with no safety driver present.*
- *The US Department of Transportation issued guidance for automated driving pilot programmes.*
- *GM and Ford set up new automated driving divisions to accelerate AV deployment. Tesla rollout Level 2 and Level2+ features with Level 3 and Level 4 features planned for 2020.*
- *Waymo, a spinoff from technology giant Google, starts first AV commercial business model in Arizona.*

Asia Pacific

- *Japan considered policies related to liabilities, driving licenses and cybersecurity laws.*
- *In China, 11 roads were added to the existing 33 roads in Beijing designated for autonomous driving tests. AVs were required to complete 5000 km of daily driving in designated closed test fields before being allowed to apply for public road testing permits.*
- *China granted permission to Audi, BMW and Daimler to test AVs in Beijing and Shanghai.*

So the report, generated by the manufacturer's association, is significantly more conservative in timescale than the claims of the individual manufacturers.

However, the report goes on to predict that "...... From 2021 onwards, some early generation Level 4 automation features may be introduced. These could include highly automated highway pilot, automated valet parking and automated vehicles such as taxis operating within virtually defined or 'geofenced' zones in urban areas.

. . . .

. . . . *ECONOMIC IMPACT OF CAVS ON THE UK ECONOMY BY 2030*

. . . . . . *a reduction of over 47,000 in serious collisions and a further 3,900 lives saved. According to current projections, a total of almost 56,000 crashes of all kinds can be eliminated via AV related technologies by 2030 alone. These are expected to be further augmented in years to come when V2X applications, such as intersection collision warning and road hazard warning, complement the safety benefits that automation will bring.*

CREATION OF MORE THAN 420,000 JOBS ACROSS THE ECONOMY

Digitisation of the automotive value chain is forecast to help create more than 20,000 new jobs in the automotive sector alone. Of these, 11,000 (55%) are expected to be highly skilled across both upstream and downstream services. Rewarding new job opportunities are expected to emerge in software and hardware development for automated and connected technologies in the upstream, and in vehicle fleet and network management in the downstream along the value chain.

Over the next decade, testing, validation and digital technology-based jobs are likely to enjoy significant growth, helping to offset changes elsewhere in traditional manufacturing and production. Current CAV development trends indicate that market-specific validation and testing will be critical for successful deployment of CAVs and this will lead to emergence of a new set of automotive jobs in vehicle system testing in the UK that does not exist currently.

The wider impact on the UK job sector within adjacent industries, including in telecommunications, content creation, logistics and others, is likely to be even more pronounced; more than 400,000 new jobs are expected to be created. These assessments are predicated on the UK leaving the EU with a favourable Brexit deal that maintains the £145 billion economic impact by 2035, instead of the currently forecast 2040.

. . . . *However, all of this will only be possible with active and sustained support from the government, especially in terms of investment in infrastructure and regulatory support.* (Source: The UK Society of Motor Manufacturers and Traders (SMMT) issued a Report "Connected and Autonomous Vehicles 2019 Report / Winning the Global Race to Market. © 2019, SMMT.)"

The UK newspaper, *The Guardian*, reported that, at the launch of the report:

Mike Hawes, the chief executive of the SMMT, said more than £500 m had been invested in research and development by industry and government, and another £740m in communications infrastructure to enable autonomous cars to work.

He said: 'The opportunities are dramatic – new jobs, economic growth and improvements across society'.

Hawes said widespread trials of autonomous vehicles were already under way and the industry would attempt to achieve the government's ambition of driverless vehicles on the roads by 2021. (Source: Gwyn Topham, Self-driving cars could provide £62bn boost to UK economy by 2030, April, © 2019, Guardian News & Media Limited.)

The French government has stated its support for the development of automated vehicles, with the aim of deploying 'highly automated' vehicles on public roads between 2020 and 2022. The government has made 40 million euros ($46 million) available to help subsidise new projects. According to Automotive News Europe (2019), more than 50 autonomous-vehicle test projects have taken place in France since 2014, including robotaxis, buses and private vehicles.

French president Emmanuel Macron has appointed a senior official, Anne-Marie Idrac, to develop a national strategy for driverless mobility – including new laws, regulations for experiments and pilot projects, and cybersecurity and privacy issues. France's Transport Minister Élisabeth Borne has announced a plan to develop transit with autonomous electric vehicles

According to the Europe Autonews report, 'the first legal proposals are expected by the end of this year and once approved will allow Level 3 and 4 passenger vehicles, driverless mass transit such as robotaxis, and automated delivery vehicles'.

Small, driverless shuttle buses built by Nayva – which has financial backing from French supplier Valeo, among others – have been operating on closed road circuits in La Defense, the business zone outside Paris, since 2017. (Source: Peter Sigal, France pushes for 'highly automated' vehicles by 2022, August 08, © 2018, Crain Communications, Inc.)

A contact in the Conseil Général de l'Environnement et du Développement Durable advised that a proposal called LOM (la loi d'orientation des mobilités or, mobility orientation law) has already made it through the senate. It will allow autonomous shuttles to circulate freely from 2020 on the entire public road network.

Germany enacted the 'Autonomous Vehicle Bill' in June 2017, modifying the existing Road Traffic Act defining the requirements for highly and fully automated vehicles, while also addressing the rights of the driver. The legislation will define what an automated vehicle is and states that such technology must comply with traffic regulations, recognise when the driver needs to resume control, and inform him or her with sufficient lead time as well as at any time permitting the driver to manually override or deactivate the automated driving mode.

In highly decentralised Germany, autonomous testing legislation is handed out by city regulatory authorities, rather than central government. Several sources advise that the current federal government plans to create an infrastructure suitable for Level 5 fully autonomous vehicles by the end of the current legislative period.

Looking behind the marketing hype of the auto manufacturers, it would seem to be reasonable to summarise that a major of auto manufacturers are focussing on the partial automation of high-end cars with a short-term objective of 2020, though 2021–2022 may be a more achievable goal for most manufacturers. These are probably about the earliest years that Level 4 automation is most likely to be available.

At Level 3 automation, the driver will still need to be alert and ready to 'immediately' take over control if the driving system encounters a situation it cannot resolve. In a situation where the driver is not driving for considerable periods of time, just how long it takes for a driver to be able to take over and recover what is probably an emergency stop or avoidance manoeuvre, remains the subject of further research and debate. Despite the hype of the motor manufacturers websites, and their claims at motor shows, in reality, full automation is expected to debut much later – but whether that is the 2020s, 2030s or 2040s is still under debate.

That said, the car manufacturer Tesla, already a pioneer in deployed degrees of automation, continues to press ahead (2019).

All new Tesla cars come standard with advanced hardware capable of providing Autopilot features today, and full self-driving capabilities in the future—through software updates designed to improve functionality over time.

Advanced Sensor Coverage

Eight surround cameras provide 360 degrees of visibility around the car at up to 250 meters of range. Twelve updated ultrasonic sensors complement this vision, allowing for detection of both hard and soft objects at nearly twice the distance of the prior system. A forward-facing radar with enhanced processing provides additional data about the world on a redundant wavelength that is able to see through heavy rain, fog, dust and even the car ahead.

Processing Power Increased 40x

To make sense of all of this data, a new onboard computer with over 40 times the computing power of the previous generation runs the new Tesla-developed neural net for vision, sonar and radar processing software. Together, this system provides a view of the world that a driver alone cannot access, seeing in every direction simultaneously, and on wavelengths that go far beyond the human senses.

Tesla Vision

To make use of a camera suite this powerful, the new hardware introduces an entirely new and powerful set of vision processing tools developed by Tesla. Built on a deep neural network, Tesla Vision deconstructs the car's environment at greater levels of reliability than those achievable with classical vision processing techniques.

Autopilot

Autopilot advanced safety and convenience features are designed to assist you with the most burdensome parts of driving. Autopilot introduces new features and improves existing functionality to make your Tesla safer and more capable over time. Autopilot enables your car to steer, accelerate and brake automatically within its lane.

Navigate on Autopilot

Navigate on Autopilot suggests lane changes to optimize your route, and makes adjustments so you don't get stuck behind slow cars or trucks. When active, Navigate on Autopilot will also automatically steer your vehicle toward highway interchanges and exits based on your destination.

Current Autopilot features require active driver supervision and do not make the vehicle autonomous.

Full Self-Driving Capability

All new Tesla cars have the hardware needed in the future for full self-driving in almost all circumstances. The system is designed to be able to conduct short and long distance trips with no action required by the person in the driver's seat.

All you will need to do is get in and tell your car where to go. If you don't say anything, the car will look at your calendar and take you there as the assumed destination or just home if nothing is on the calendar. Your Tesla will figure out the optimal route, navigate urban streets (even without lane markings), manage complex intersections with traffic lights, stop signs and roundabouts, and handle densely packed freeways with cars moving at high speed. When you arrive at your destination, simply step out at the entrance and your car will enter park seek mode, automatically search for a spot and park itself. A tap on your phone summons it back to you.

Some features require turn signals and are limited in range. The future use of these features without supervision is dependent on achieving reliability far in excess of human drivers as demonstrated by billions of miles of experience, as well as regulatory approval, which may take longer in some jurisdictions. As these self-driving capabilities are introduced, your car will be continuously upgraded through over-the-air software updates. (Source: Tesla.com, Future of Driving. © 2019, Tesla.)

In a podcast interview with the money management firm ARK Invest, Elon Musk, the Tesla CEO, made a prediction that the Tesla's full self-driving feature will be completed by the end of 2019. And by the end of 2020, he added, it will be so capable, you'll be able to snooze in the driver's seat while it takes you from your parking lot to wherever you're going.

> I think we will be "feature-complete" on full self-driving this year, meaning the car will be able to find you in a parking lot, pick you up, take you all the way to your destination without an intervention this year, I am certain of that. That is not a question mark.

Today, for an extra $5000 at purchase, owners can unlock their vehicles' 'enhanced autopilot feature'. The technology 'guides a car from a highway's on-ramp to off-ramp, including suggesting and making lane changes, navigating highway interchanges, and taking exits', according to the driver's manual for the S and X models.

However, Tesla has had to send notification to all owners, and add into every Tesla drivers manual *'Traffic-Aware Cruise Control cannot detect all objects and may not brake/decelerate for stationary vehicles, especially in situations when you are driving over 50 mph (80 km/h) and a vehicle you are following moves out of your driving path and a stationary vehicle or object is in front of you instead'.*

A stark and basic warning, made as the result of several crashes (Figure 1.5).

Figure 1.5 Screenclip from press report. Source: Dailymail.co.uk.

Figure 1.6 Publicly available EC Document from EC project Ensemble. Source: ENSEMBLE, ENabling SafE Multi-Brand pLatooning for Europe, European Union, 2017. Licensed under CC By 4.0.

Moving attention from cars to large goods vehicles, there has been much attention to, research and investment in truck platooning. Truck platooning is the linking of two or more trucks in convoy, using connectivity technology and automated driving support systems. Much as with adaptive cruise control, but with communication between vehicles, rather than just sensors, these vehicles automatically maintain a set, close distance between each other when they are connected for certain parts of a journey, for instance on motorways (Figure 1.6).

The truck at the head of the platoon acts as the leader, with the vehicles behind reacting and adapting to changes in its movement – requiring little to no action from drivers. In the first instance, drivers will remain in control at all times, so they can also decide to leave the platoon and drive independently. However, according to Commercial Fleet News, and after investing some €50 million (£44 m) on platoon testing,

> *Mercedes-Benz Trucks has concluded that there is no business case for truck platooning, saying that the technology failed to deliver appreciable fuel savings in its on-the-road tests'.* (2019)
>
> *Although the manufacturer will remain committed to ongoing platooning projects with partners, such as Ensemble in Europe, it now plans to refocus its resources on developing autonomous, self-driving technologies in its trucks.*
>
> *It told delegates at this month's Consumer Electronics Show (CES) in Las Vegas that results show fuel savings, even in perfect platooning conditions, were less than expected. Savings were further diminished when the platoon was disconnected and the trucks had to accelerate to reconnect. In at least four US long-distance applications, analysis showed no business case for driving platoons with new, aerodynamic trucks.* (Source: Commercialfleet.com, Mercedes switches focus away from platooning trials. © 2019, Bauer Consumer Media.)

Congestion and the nature of the UK and many European countries networks have always been highlighted as possible barriers to vehicle platooning working effectively/realising benefits in Europe.

While platooning looks good as a concept on paper, in the real world it is weighed down by practical issues like pairing trucks from different fleets with incompatible hardware and software, and Mercedes experience shows that it is difficult to actually realise the apparent benefits, such as fuel savings. Other manufacturers are likely to turn their attention to the longer term goals of fully automated driving as fleet operator interest in its potential savings declines.

Vehicles that drive themselves could bring dramatic shifts in car ownership, public transport, employment patterns, business and urban development.

The theoretical safety benefits are huge. 'Autonomous' viz. 'automated' vehicles won't drink and drive or get distracted by telephone calls, Facebook posts, or children in the back. They will be programmed to drive at appropriate and legal speeds, and will pay attention to their environment in 360 degrees at millions of times every second.

Whether for cars, vans or large commercial vehicles, the adoption and take up of these technologies will mitigate some risks; but it is likely that they may also create new risks. And there are still many operational, research and regulatory questions that partly automated and fully automated vehicles present, that remain unsolved.

2

Automated Driving Levels

2.1 SAE J3016

Following several inconsistent attempts to classify the different types and levels of automation (examples: NHTSA, 2013; BAST – Gasser, 2012; VDA, 2015; and several others), the International Society of Automotive Engineers (SAE) delivered a harmonised classification system for Automated Driving Systems (ADS), in its publication *Recommended Practice: SAE J3016: Taxonomy and Definitions for Terms Related to On-Road Motor Vehicle Automated Driving Systems* (SAE International, 2016). SAE J3016 was first issued in January 2014 and revised in September 2016, notably after input from the United States–based CAMP Projected (Crash Avoidance Metrics Partnership).

This taxonomy has been widely adopted, including by the European Commission in its C-ITS Platform Report (2016) and its C-Roads programme; by US DoT, ISO TC204 Intelligent Transport Systems, ISO TC22, Road Vehicles, and CEN TC278 *Intelligent Transport Systems*, and is in use also by UNECE WP29 World Forum for Harmonisation of Vehicle Regulations 'Revised Framework document on automated/autonomous vehicles' (2019).

This chapter recommends the use of J3016 taxonomy, and explains its key features, adequate for the task of this publication whose purpose is to identify, assess, and suggest strategies to overcome the barriers that face the implementation of automated driving and Mobility as a Service (MaaS), where the use of a clear and structured taxonomy is essential to understanding the issues. However this author strongly recommends that anyone who has more than a layman's interest in these subjects connects to SAE at www .sae.org/standards/content/J3016 201 806 (or simply google SAE J3016), where you are able to download a free .pdf of J3016.

While this chapter describes the key features of J3016, J3016 contains many examples of the subtlety of the automated driving paradigm as well as many definitions and discussion that are very helpful to understand the subtleties of this subject. While the author has made best efforts to interpret and extract the aspects relevant to the subject of this publication, there is no substitute to reading the original published (and free) issue of this well-elaborated document, and I recommend the reader to do so.

SAE J3016 provides a taxonomy describing the full range of levels of driving automation in on-road motor vehicles (see Figure 2.1). These levels primarily identify whether it is the human or the machine in charge of what J3016 calls 'the dynamic driving task' (DDT). This ranges from Level 0 where the dynamic driving task is entirely performed by the human driver (no automation) to Level 5 where the dynamic driving task is entirely

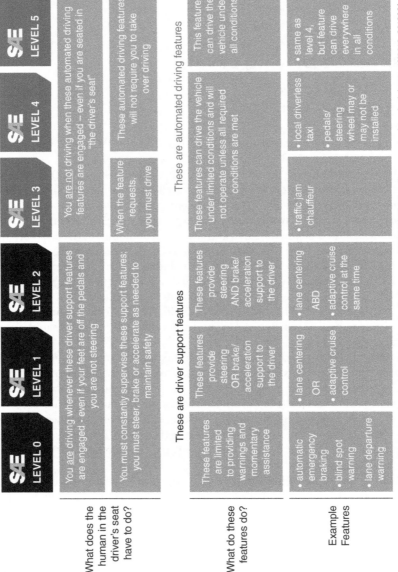

SAE J3016™ LEVELS OF DRIVING AUTOMATION

	SAE LEVEL 0	SAE LEVEL 1	SAE LEVEL 2	SAE LEVEL 3	SAE LEVEL 4	SAE LEVEL 5
What does the human in the driver's seat have to do?	You are driving whenever these driver support features are engaged - even if your feet are off the pedals and you are not steering			You are not driving when these automated driving features are engaged – even if you are seated in "the driver's seat"		
	You must constantly supervise these support features; you must steer, brake or accelerate as needed to maintain safety			When the feature requests, you must drive	These automated driving features will not require you to take over driving	
		These are driver support features		These are automated driving features		
What do these features do?	These features are limited to providing warnings and momentary assistance	These features provide steering OR brake/ acceleration support to the driver	These features provide steering AND brake/ acceleration support to the driver	These features can drive the vehicle under limited conditions and will not operate unless all required conditions are met		This feature can drive the vehicle under all conditions
Example Features	• automatic emergency braking • blind spot warning • lane departure warning	• lane centering OR • adaptive cruise control	• lane centering AND • adaptive cruise control at the same time	• traffic jam chauffeur	• local driverless taxi • pedals/ steering wheel may or may not be installed	• same as level 4, but feature can drive everywhere in all conditions

For a more complete description, please download a free copy of SAE J3016: https://www.sae.org/standards/content/J3016_201806/

Figure 2.1 SAE J3016 freely available document. Source: Jennifer Shuttleworth, SAE Standards News: J3016 automated-driving graphic update. © 2019, SAE International.

Table 2.1 Summary of levels of driving automation.

				dynamic driving task			
LEVEL	Name	Narrative definition	Sustained lateral and longitudinal vehicle motion control	Object and Event Detection and Response	Direct Driving Task fallback	Operational Design Domain
		Driver performs part or all of the Direct Driving Task				
0	**No Driving Automation**	The performance by the *driver* of the entire *Direct Driving Task*, even when enhanced by *active safety systems*.	*Driver*	*Driver*	*Driver*	*n/a*
1	**Driver Assistance**	The *sustained* and *Operational Design Domain*–specific execution by a *driving automation system* of either the *lateral* or the *longitudinal vehicle motion control* subtask of the Direct Driving Task (but not both simultaneously) with the expectation that the *driver* performs the remainder of the *Direct Driving Task.*	*Driver* and *System*	*Driver*	*Driver*	*Limited*
2	**Partial Driving Automation**	The *sustained* and *Operational Design Domain*–specific execution by a *driving automation system* of both the *lateral* and *longitudinal vehicle motion control* subtasks of the *Direct Driving Task* with the expectation that the *driver* completes the *Object and Event Detection and Response* subtask and *supervises* the *driving automation system.*	*System*	*Driver*	*Driver*	*Limited*

(Continued)

Table 2.1 (Continued)

			dynamic driving task		
					Fallback-ready (user becomes the driver during fallback)	
		Automated Driving System ('System') performs the entire Direct Driving Task (while engaged)				
3	**Conditional Driving Automation**	The sustained and Operational Design Domain–specific performance by an Automated Driving System of the entire Direct Driving Task with the expectation that the Direct Driving Task fallback-ready user is receptive to automated Driving System–issued requests to intervene, as well as to Direct Driving Task performance relevant system failures in other vehicle systems,	System	System	Fallback-ready (user becomes the driver during fallback)	Limited
4	**High Driving Automation**	The *sustained* and *Operational Design Domain–*specific performance by an *Automated Driving System* of the entire *Direct Driving Task* and *Direct Driving Task fallback* without any expectation that a *user* will respond to a *request to intervene.*	System	System	System	Limited
5	**Full Driving Automation**	The sustained and unconditional (i.e. not Operational Design Domain specific) performance by an Automated Driving System of the entire Direct Driving Task and Direct Driving Task fallback without any expectation that a System of the entire Direct Driving Task and Direct Driving Task fallback without any expectation that a user will respond to a request to intervene.	System	System	System	Unlimited

Source: SAE Standards, SURFACE VEHICLE RECOMMENDED PRACTICE J3016™ SEP 2016 IJ3016. © 2016, SAE International. Source: Modified from J3016

performed by the automated driving system (full automation). The intermediate levels represent a shared performance of the dynamic driving task between the driver and the vehicle, either in a simultaneous or in a sequential way.

The dynamic driving task comprises both the lateral control (steering) and the longitudinal control (accelerating, braking) of the vehicle, together with the monitoring of the environment, referred to in J3016 as the 'object and event detection and response' (OEDR).

Confusingly, J3016 uses both the terms 'driving automation system' (DAS) and 'automated driving system' (ADS).

In J3016, 'driving automation system' (DAS) means 'The hardware and software that are collectively capable of performing part or all of the dynamic driving task on a sustained basis; this term is used generically to describe any system capable of level 1–5 driving automation', whereas 'automated driving system' (ADS) features (i.e. Levels 3–5) that perform the complete *dynamic driving task*, including crash avoidance capability.

As specified in SAE J3016, *'these levels are descriptive rather than normative and technical rather than legal. They imply no particular order of market introduction. Elements indicate minimum rather than maximum system capabilities for each level. A particular vehicle may have multiple driving automation features such that it could operate at different levels depending upon the feature(s) that are engaged'.*

My complaint here is not the segregation, but the similarity of terminology, which is not clear without careful reading of the detail. DAS would be better described as 'driver assistance systems' or preferably a different acronym, such as 'driver support systems'(DSS) or something that segregated that in one case there was a driver in the vehicle in control, and in the other not. However, J3016 is published so this is water under the bridge.

SAE International recently unveiled a new visual chart (below) that is designed to clarify and simplify its J3016 'levels of driving automation' standard for consumers. The J3016 standard defines six levels of driving automation, from SAE Level 0 (no automation) to SAE Level 5 (full vehicle autonomy). It serves as the industry's most-cited reference for automated-vehicle (AV) capabilities.

The update is a recent iteration of the J3016 graphic first deployed in 2016. As the industry gets closer to producing AVs in volume, the SAE J3016 Technical Standards Committee saw the need to more clearly explain the features in each of the six driving levels, and how they relate to consumers' increased safety and convenience, noted Jack Pokrzywa, SAE's Ground Vehicle Standards Director.

The latest J3016 graphic is a living document. It will continue to evolve gradually as the industry and the technical standard J3016 itself evolves. It is freely available from the website shown at the foot of the graphic.

The SAE recommended practice describes in some detail, what it calls 'motor vehicle driving automation systems that perform part or all of the dynamic driving task (DDT) on a sustained basis'.

J3016 provides a taxonomy with detailed definitions for six levels of driving automation, ranging from *no* driving automation (Level 0) to *full* driving automation (Level 5), in the context of *motor vehicles* and their *operation* on roadways.

The definitions are intended to be appropriate for all categories of vehicles (not just cars), including, it states '... as well as motorcyclists, pedal cyclists, and pedestrians' (although the concept of an 'automated' pedestrian, is quite novel . . . . Perhaps robotics is advancing faster than I have noticed and they are envisaging the domestic robot running around the streets doing grannies shopping!).

The levels apply to the *driving automation feature(s)* that are engaged in any given instance of on-road operation of an equipped vehicle. As such, although a given *vehicle* may be equipped with a driving automation system that is capable of delivering multiple driving automation features that perform at different levels, the level of driving automation exhibited in any given instance is determined by the *feature(s)* that are engaged.

J3016 also refers to roles of three primary actors in driving: the (human) *user*, the *driving automation system*, and other *vehicle* systems and components, 'role' in this context refers to the expected role of a given primary actor, based on the design of the driving automation system in question and not necessarily to the actual performance of a given primary actor. For example, a driver who fails to *monitor* the roadway during engagement of a Level 1 adaptive cruise control (ACC) system still has the role of *driver*, even while (s)he is neglecting it.

J3016 also provides clear definitions of the terms used in the document.

J3016 excludes active safety systems, such as electronic stability control and automated emergency braking, and certain types of driver assistance systems, such as lane keeping assistance, etc., from its driving automation taxonomy because they do not perform part or all of the dynamic driving task on a sustained basis and, rather, merely provide momentary intervention during potentially hazardous situations. Due to the momentary nature of the actions of active safety systems, J3016 contends that their intervention does not change or eliminate the role of the driver in performing part or all of the dynamic driving task, and thus are not considered to be driving automation.

However, in J3016, crash avoidance features, including intervention-type active safety systems, may be included in vehicles equipped with driving automation systems at any level. For automated driving system (ADS) features (i.e. Levels 3–5) that perform the complete 4dynamic driving task, crash avoidance capability is part of ADS functionality.

To understand the apparent duplication of terms sore the same direct driving task (DAS and ADS), as explained above, it is worth reminding the reader that J3016 defines a 'driving automation system or technology' (DAS/DAT) as *'The hardware and software that are collectively capable of performing part or all of the dynamic driving task on a sustained basis; this term is used generically to describe any system capable of level 1–5 driving automation'.*

In contrast to this generic term for any Level 1–5 system, J3016 defines the specific term for a Level 3–5 system as 'automated driving system (ADS)' (Figure 2.2).

J3016 freely provides descriptions of a wide range of use cases of fallback situations before Level 5, where the driver has to take back control of driving the vehicle. At Level 1 (example, ACC), this is not so challenging because the driver is still steering the vehicle and is watching out for threats from other vehicles.

But at Level 4 – even Level 3, the user will not be operating the vehicle in real time: He/ she may well be texting/doing email, reading a newspaper or book, etc. These activities are legitimate, and probably the reason the driver elected to use automated control. But, whereas in the ACC situation, the driver is ready to resume control quickly – at any time – well within a second or so, In the case of L3 and L4 there is active discussion as to how long it will take for a vehicle user to (physically and mentally) take back control. There are many theories/ but it could be typically 8+ seconds. In a potential crash situation, there is not normally 3–4 seconds advance warning time.

The SAE publication (J3016) determines six (0–5) discrete and mutually exclusive levels of *driving automation* (see its 'levels of driving automation' chart reproduced near the start of this chapter). Central to this taxonomy are the respective roles of the

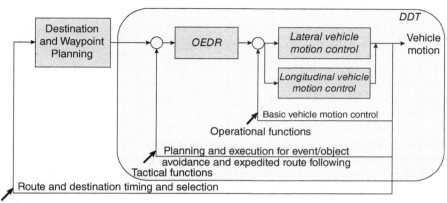

Figure 2.2 J3016 schematic (not a control diagram) view of driving task showing dynamic driving task portion. Source: SAE Standards, SURFACE VEHICLE RECOMMENDED PRACTICE J3016™ SEP 2016 IJ3016. © 2016, SAE International.

(human) *user* and the *driving automation system* in relation to each other. Because changes in the functionality of a *driving automation system* change the role of the (human) *user*, they provide a basis for categorising such *system features*.

For example, if the *driving automation system* performs the *sustained longitudinal* and/or *lateral vehicle motion control* subtasks of the *dynamic driving task*, the *driver* does not do so, although (s)he is expected to complete the *dynamic driving task*. This division of roles corresponds to Levels 1 and 2.

If the *driving automation system* performs the entire *dynamic driving task*, the *user* does not do so. However, if a *dynamic driving task fallback-ready user* is expected to take over the *dynamic driving task* when a *dynamic driving task performance-relevant system failure* occurs or when the *driving automation system* is about to leave its *operational design domain (ODD)*, then that *user* is expected to be *receptive* and able to resume *dynamic driving task* performance when alerted to the need to do so. This division of roles corresponds to Level 3.

Lastly, if a *driving automation system* can perform the entire *dynamic driving task* and *dynamic driving task fallback* either within a prescribed *operational design domain* or in all driver-manageable on-road driving situations (unlimited operational design domain), then any *users* present in the *vehicle* while the *automated driving system* is engaged are *passengers*. This division of roles corresponds to Levels 4 and 5.

One area of ambivalence in J3016 that requires clarification, and better definition in regulations, is the dichotomy between the J3016 levels of automation chart, which states for Level 4 automation: 'The sustained and Operational Design Domain – specific performance by an Automated Driving System of the entire Direct Driving Task and Direct Driving Task fallback *without any expectation that a user will respond to a request to intervene*' and the summary chart, which consistently states 'Pedals/steering wheel may or may not be installed', and the explanatory detailed text in J3016, which states:

'*Sample use case sequence at Level 4 showing Automated Driving System engaged and occurrence of an Automated Driving System failure that does not prevent continued dynamic driving task performance by an available human user. The Automated Driving System feature may prompt a passenger seated in the driver's seat (if available) to resume dynamic driving task performance; if no driver's seat with receptive passenger, the Automated Driving System automatically achieves a minimal risk condition.'*

Either Level 4 are limited use cases of full automation, in which case prompting a passenger is inappropriate or it is a Level 3.5. Clarification is needed.

The *vehicle* also fulfils a role in this *driving automation* taxonomy, but the role of the *vehicle* does not change the role of the *user* in performing the *dynamic driving task*.

In this way J3016 defines *driving automation systems* are categorised into levels based on:

a. Whether the driving automation system performs *either* the longitudinal <u>or</u> the lateral vehicle motion control subtask of the dynamic driving task.
b. Whether the *driving automation system* performs *both* the *longitudinal* <u>and</u> the *lateral vehicle motion control* subtasks of the *dynamic driving task* simultaneously.
c. Whether the *driving automation system* also performs the *OEDR* subtask of the *dynamic driving task*.
d. Whether the *driving automation system* also performs *dynamic driving task* fallback.
e. Whether the *driving automation system* is limited by an *Operational Design Domain*.

SAE's describes its levels of *driving automation* as descriptive and informative, rather than normative, and technical rather than legal. The table below summarises the six levels of *driving automation* in terms of these five elements.

The roles of the ***driver*** and the ***Driving Automation System*** can therefore be determined as:

Level 0 – No Driving Automation
In this paradigm:
The ***driver*** (at all times):

- Performs the entire dynamic driving task

The ***Driving Automation System*** (if any):

- Does not perform any part of the dynamic driving task on a sustained basis (although other vehicle systems may provide warnings or support, such as momentary emergency intervention)

Level 1 – Driver Assistance
In this paradigm:
The ***driver*** (at all times):

- Performs the remainder of the dynamic driving task not performed by the driving automation system
- Supervises the driving automation system and intervenes as necessary to maintain safe operation of the vehicle
- Determines whether/when engagement or disengagement of the driving automation system is appropriate
- Immediately performs the entire dynamic driving task whenever required or desired

The ***Driving Automation System*** (while engaged):

- Performs part of the dynamic driving task by executing either the longitudinal or the lateral vehicle motion control subtask
- Disengages immediately upon driver request

Level 2 – Partial Driving Automation
The *Driver* (at all times):

- Performs the remainder of the dynamic driving task not performed by the driving automation system
- Supervises the driving automation system and intervenes as necessary to maintain safe operation of the vehicle
- Determines whether/when engagement and disengagement of the driving automation system is appropriate
- Immediately performs the entire dynamic driving task whenever required or desired

The *Driving Automation System* (while engaged):

- Performs part of the dynamic driving task by executing both the lateral and the longitudinal vehicle motion control subtasks
- Disengages immediately upon driver request

Level 3 – Conditional Driving Automation
The *Driver* (while the Automated Driving System is not engaged):

- Verifies operational readiness of the Automated Driving System–equipped vehicle
- Determines when engagement of Automated Driving System is appropriate
- Becomes the Direct Driving Task fallback-ready user when the Automated Driving System is engaged

The *Direct Driving Task fallback-ready user* (while the Automated Driving System is engaged):

- Is receptive to a request to intervene and responds by performing dynamic driving task fallback in a timely manner
- Is receptive to dynamic driving task performance-relevant system failures in vehicle systems and, upon occurrence, performs dynamic driving task fallback in a timely manner
- Determines whether and how to achieve a minimal risk condition
- Becomes the driver upon requesting disengagement of the Automated Driving System

The *Automated Driving System* (while engaged):
Automated Driving System (while not engaged):

- Permits engagement only within its Operational Design Domain

Automated Driving System (while engaged):

- Performs the entire Direct Driving Task
- Determines whether Operational Design Domain limits are about to be exceeded and, if so, issues a timely request to intervene to the Direct Driving Task fallback-ready user
- Determines whether there is a Direct Driving Task performance-relevant system failure of the Automated Driving System and, if so, issues a timely request to intervene to the Direct Driving Task fallback-ready user
- Disengages an appropriate time after issuing a request to intervene
- Disengages immediately upon driver request

Level 4 – High Driving Automation
The ***Driver/dispatcher*** (while the Automated Driving System is not engaged):

- Verifies operational readiness of the Automated Driving System–equipped vehicle
- Determines whether to engage the Automated Driving System
- Becomes a passenger when the Automated Driving System is engaged only if physically present in the vehicle

The ***Passenger/dispatcher*** (while the Automated Driving System is engaged):

- Need not perform the Direct Driving Task or Direct Driving Task fallback
- Need not determine whether and how to achieve a minimal risk condition
- May perform the Direct Driving Task fallback following a request to intervene
- May request that the Automated Driving System disengage and may achieve a minimal risk condition after it is disengaged
- May become the driver after a requested disengagement

The ***Automated Driving System*** (while not engaged):

- Permits engagement only within its Operational Design Domain

The ***Automated Driving System*** (while engaged):

- Performs the entire Direct Driving Task
- May issue a timely request to intervene
- Performs Direct Driving Task fallback and transitions automatically to a minimal risk condition when:
  - A Direct Driving Task performance-relevant system failure occurs or
  - A user does not respond to a request to intervene or
  - A user requests that it achieve a minimal risk condition
- Disengages, if appropriate, only after:
  - It achieves a minimal risk condition or
  - A driver is performing the Direct Driving Task
- May delay user-requested disengagement

Level 5 – Full Driving Automation
The ***Driver/dispatcher*** (while the Automated Driving System is not engaged):

- Verifies operational readiness of the Automated Driving System–equipped vehicle
- Determines whether to engage the Automated Driving System
- Becomes a passenger when the Automated Driving System is engaged only if physically present in the vehicle

The ***Passenger/dispatcher*** (while the Automated Driving System is engaged):

- Need not perform the Direct Driving Task or Direct Driving Task fallback
- Need not determine whether and how to achieve a minimal risk condition

The ***Automated Driving System*** (while not engaged):

- Permits engagement of the Automated Driving System under all driver-manageable on-road conditions

The ***Automated Driving System*** (while engaged):

- Performs the entire Direct Driving Task

- Performs Direct Driving Task fallback and transitions automatically to a minimal risk condition when:
- A Direct Driving Task performance-relevant system failure occurs or
- A user does not respond to a request to intervene or
- A user requests that it achieve a minimal risk condition
- Disengages, if appropriate, only after:
- It achieves a minimal risk condition or
- A driver is performing the Direct Driving Task
- May delay a user-requested disengagement

What J3016 does not appear to cover very well is the paradigm of an automated vehicle in a MaaS situation. As can be seen above, in Level 4 automation, the passenger/despatcher 'May become the driver after a requested disengagement'.

But in the MaaS paradigm the passenger may not be a qualified driver. So J3016 assumes that in these circumstances the despatcher will be ready to take over and drive the vehicles remotely. The author is not confident that this will always be practicable. In reality, a MaaS despatcher may be handing many vehicles simultaneously and may not be available to remotely drive the vehicle. Similarly, in a situation where Level 4 automated mode has been selected, it cannot be assumed that there will always be a qualified driver in the vehicle capable to take over the driving task, or (s)he may be over the alcohol limit or asleep. A Level 5 system may have no physical driver control equipment. Further work needs to be undertaken concerning this paradigm.

The following table describes a *user's* role with respect to an engaged *driving automation system* operating at a particular level of *driving automation* at a particular point in time (Table 2.2).

A *user* occupying a given *vehicle* can have one of three possible roles during a particular *trip*: (i) *driver*, (ii) *dynamic driving task fallback-ready user* or (iii) *passenger*. A *remote user* of a given *vehicle* (i.e. who is not seated in the driver's seat of the *vehicle* during use) can also have one of three possible roles during a particular *trip*: (i) *remote driver*, (ii) *dynamic driving task fallback-ready user* or (iii) *driverless operation dispatcher*.

Note: A *vehicle* equipped with a Level 4 or 5 *Automated Driving System* may also support a *driver* role. For example, in order to complete a given *trip*, a *user* of a *vehicle* equipped with a Level 4 *Automated Driving System* feature designed to *operate* the *vehicle* during high-speed freeway conditions will generally choose to perform the *dynamic driving task* when the freeway ends; otherwise the *Automated Driving System* will automatically perform *dynamic driving task fallback* and achieve a *minimal risk condition* as needed. However, unlike at Level 3, this *user* is not a *dynamic driving task fallback-ready user* while the *Automated Driving System* is engaged.

Table 2.2 User roles while a driving automation system is engaged.

No *Driving Automation*	Engaged Level of *Driving Automation*				
0	1	2	3	4	5
In-vehicle user	*Driver*		Dynamic driving task	*Passenger*	
Remote User	*Remote Driver*		*Fallback ready user*	*Driverless operation dispatcher*	

Source: SAE Standards, SURFACE VEHICLE RECOMMENDED PRACTICE J3016™ SEP 2016 IJ3016. © 2016, SAE International.

2.2 The Significance of Operational Design Domain (ODD)

J3016 explains that Levels 1 through 4 expressly contemplate *operational design domain* limitations. In contrast, Level 5 does not normally have *operational design domain* limitations.

Accordingly, it goes on to say that accurately describing a *feature* (other than at Level 5) requires identifying both its level of *driving automation* and its *ODD*. This combination of level of *driving automation* and *operational design domain* is called a *usage specification*, and a given *feature* satisfies a given *usage specification*.

Because of the wide range of possible *operational design domain*s, a wide range of possible *features* may exist in each level (e.g. Level 4 includes parking, high-speed, low-speed, geo-fenced). For this reason, SAE J3016 provides less detail about the *operational design domain* attributes that may define a given *feature* than about the respective roles of a *driving automation system* and its *user*.

Operational design domain is especially important to understanding why a given *automated driving system* is not Level 5 merely because it *operates* an *automated driving system–dedicated vehicle*. Unlike a Level 5 *automated driving system*, a Level 4 *automated driving system* has a limited *operational design domain*. Geographic or environmental restrictions on an *automated driving system – DV* may reflect the *operational design domain* limitations of its *automated driving system* (or they may reflect *vehicle* design limitations).

J3016 explains that Level 1 to 4 *features* are subject to limited *operational design domains*. These limitations reflect the technological capability of the *driving automation system*. For example, Level 4 *automated driving system – dedicated vehicles* that operate in enclosed courses have existed for many decades as people movers and airport shuttles. The *operational design domain* for such *vehicles* is very simple, well-controlled, and physically enclosed (*vehicle* operates on a fixed course; physical barriers prevent encroachment; protected from external events, weather, etc.). This highly structured and simple *operational design domain* makes it technologically less challenging to achieve Level 4 *driving automation*.

And J3016 points out that, a Level 3 *automated driving system* feature that operates a *vehicle* on open roads in mixed traffic, and does so in environments that include inclement weather, faces a significantly higher technological bar in terms of *automated driving system* capability by virtue of the more complex and unstructured *operational design domain*.

Operational design domains for a given *driving automation system feature* may encompasses a multiple range of parameters that define the limits of that *feature's* functional capability to *operate* in design-specified on-road environments; for example: geographical restrictions, lighting conditions, read markings, road types, traffic barriers, weather conditions, etc. So an *operational design domain* may be quite varied and multi-faceted. A *feature* will *operate* as designed only when all the *operational design domain*–defining variables satisfy design criteria.

2.3 Deprecated Terms

J3016 advises not to use the terms: autonomous, driving modes(s), self-driving, unmanned, robotic, and control; and is particularly scathing about the use of the word 'autonomous':

This usage obscures the question of whether a so-called 'autonomous vehicle' depends on communication and/or cooperation with outside entities for important functionality (such as data acquisition and collection)....... if they depend on communication and/or cooperation with outside entities, they should be considered cooperative rather than autonomous. ..... in jurisprudence, autonomy refers to the capacity for self-governance. In this sense, also, 'autonomous' is a misnomer as applied to automated driving technology, because even the most advanced Automated Driving Systems are not 'self-governing'.

The recommended practice J3016 further recommends against using terms that make *vehicles*, rather than driving, the object of automation, because doing so tends to lead to confusion between *vehicles* that can be *operated* by a (human) *driver* or by an *automated driving system* and *automated driving system – dedicated vehicles*, which are designed to be *operated* exclusively by an *automated driving system*. J3016 explains that it also fails to distinguish other forms of vehicular automation that do not involve automating part or all of the *dynamic driving task*.

J3016 goes on to explain that a given *vehicle* may be equipped with a *driving automation system* that is capable of delivering multiple *driving automation features* that *operate* at different levels; thus, the level of *driving automation* exhibited in any given instance is determined by the *feature(s)* engaged. Hence its (to this authors opinion), confusing use of both the terms 'driving automation system' and 'automated driving system'. True there is the subtle difference explained, but any benefit is to this authors opinion, more than offset by the confusion that the multiple terms cause. The names used should be much easier to differentiate.

2.4 No Relative Merit

While numbered sequentially 0 through 5, J3016 levels do not specify or imply hierarchy in terms of relative merit, technology sophistication, or order of deployment.

Thus, J3016 explains that it does not specify or imply that, for example, Level 4 is 'better' than Level 3 or Level 2.

Similarly, it recommends against attempting to further specify *driving automation features* using fractional SAE J3016 levels, such as 2.5 or 4.7. 'Qualified or fractional J3016 levels would render the meaning of the levels ambiguous by removing the clarity otherwise provided by the strict apportionment of roles between the user and the driving automation system in performance of the dynamic driving task and fallback for a given vehicle'.

2.5 Mutually Exclusive Levels

The levels in the J3016 taxonomy are intentionally discrete and mutually exclusive. As such, J3016 asserts that 'it is not logically possible for a given feature to be assigned more than a single level. For example, a low-speed driving automation feature described by the manufacturer as being capable of performing the complete dynamic driving task in dense traffic on fully access-controlled freeways cannot be both level 3 and level 4, because either it is capable of automatically performing the dynamic driving task fallback and achieving a minimal risk condition whenever needed, or it relies (at least sometimes) on the driver to respond to a request to intervene and either perform the dynamic driving task or achieve a minimal risk condition on his or her own'.

J3016 explains that it is, however, *'quite possible for a driving automation system to deliver multiple features at different levels, depending on the usage specification and/or user preferences. For example, a vehicle may be equipped with a driving automation system capable of delivering, under varying conditions, a level 1 ACC feature, a level 2 highway driving assistance feature, a level 3 freeway traffic jam feature, and a level 4 automated valet parking feature – in addition to allowing the user to operate the vehicle at level 0 with no driving automation features engaged. From the standpoint of the user, these various features engage sequentially, rather than simultaneously, even if the driving automation system makes use of much of the same underlying hardware and software technology to deliver all four driving automation features'.*

2.6 J3016 Limitations

Whilst this author is keen to advocate the use of J3016 as the standard taxonomy for levels of automation within vehicles, that recommendation is made within the context of awareness of the shortcomings of J3016.

J3016 would be a lot better if it remained simpler and clearer and did not get involved in convoluted subtleties of differences in operational design domains. It also remains unclear on the role of the occupants of the vehicle at Levels 4 and 5.

Like much work led by/driven by automotive manufacturers, J3016 is car-centric, and while it recognises a role of 'system', it does not recognise that there may/will be several interacting systems and provides its taxonomy solely from the point of view of the vehicle. It characterises the taxonomy of automation levels within the vehicle, as if the vehicle is the sole entity in an automated driving system, and it does not provide either a

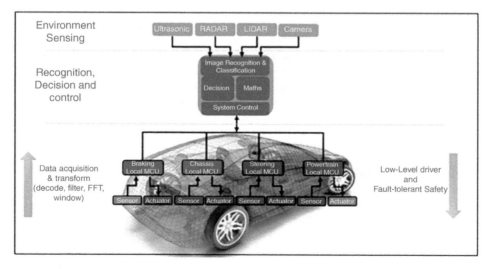

Figure 2.3 Automated car architecture Schoettle and Sivak (2015). Source: James Scobie and Mark Stachew, ELECTRONIC CONTROL SYSTEM PARTITIONING IN THE AUTONOMOUS VEHICLE, Oct 29. © 2015, European Business Press.

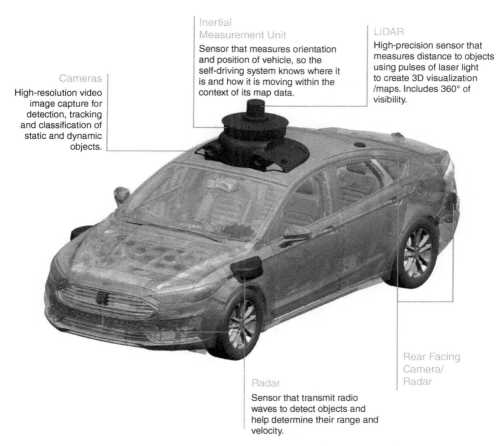

Inertial Measurement Unit

Sensor that measures orientation and position of vehicle, so the self-driving system knows where it is and how it is moving within the context of its map data.

LiDAR

High-precision sensor that measures distance to objects using pulses of laser light to create 3D visualization /maps. Includes 360° of visibility.

Cameras

High-resolution video image capture for detection, tracking and classification of static and dynamic objects.

Rear Facing Camera/ Radar

Radar

Sensor that transmit radio waves to detect objects and help determine their range and velocity.

Figure 2.4 Ford virtual driving system (eenews Automotive, 2019). Source: Ford, Matter of trust, FORD'S APPROACH TO DEVELOPING SELF-DRIVING VEHICLE, © Ford Motor Company.

description of, nor taxonomy for, the automated driving paradigm, nor provide description of the taxonomy/role of the automated vehicle within the overall road network.

Automotive manufacturers tend to envision their vehicle as the only entity in the automated driving paradigm that is in control of its operation and movement of the vehicle. These examples from EE News and Ford are typical (Figures 2.3 and 2.4).

Whereas, in reality, the vehicle (automated or not) is just one actor in a road transport network system. In instantiation, an automated driving system is better represented by the following figure, in which the vehicle is only one of 20 or more actors.

2.7 Actors in the Automated Vehicle Paradigm

The following figure shows a more realistic representation of the actors in an automated driving system. It does not purport to be a complete physical, systems or communications architecture, and is followed by a terse description of each of the actors (Figure 2.5).

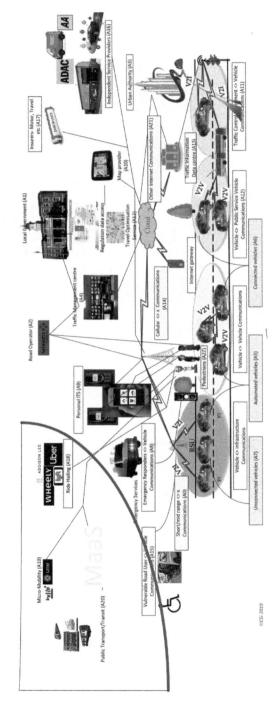

Figure 2.5 Actors in a CAV paradigm. Source: Bob Williams, Senior Consultant, CSi (UK), Oxfordshire, UK.

A complex mesh of communication and coordination, which we could show in a more structured fashion as in Figure 2.6.

In this paradigm the actors are characterised as follows:

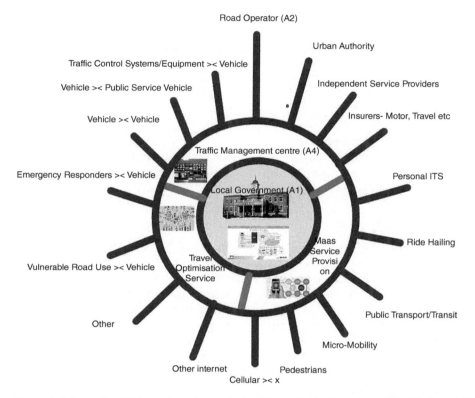

Figure 2.6 Actors in a CAV paradigm. Source: Bob Williams, Senior Consultant ,CSi (UK), Oxfordshire, UK.

2.7.1 Local Government (A1)

The local government (including enactment of national regulations) is by far the most significant actor in the automated driving paradigm. It creates (or inherits), and is the controller of, the rules, regulations and bylaws that cover each and every in./cm of the roads within its domain.

Further it has invested in and built every one of the roads within its domain (and continues to invest in maintenance and in new road building and upgrading). Further still, it provides and maintains each and every road sign. It provides all of the street furniture and equipment (traffic lights, variable message signs [except where the central government authority do this on behalf of the local authority], road signs, street lights, etc.,) and provides the police (and support officers), and traffic wardens, to enforce compliance to the rules, regulations and bylaws it has created. Finally, it operates all of the traffic management centres (TMC) that dynamically manage and control the flow of traffic on its road system.

Please understand that every automated (or so called 'autonomous') vehicle has to operate within the rules, regulations, bylaws and control systems of the local government, and operate under the dynamic control of its traffic lights, and transport optimisation service(s) and traffic instructions. So the automated vehicle is effectively controlled by the local government, and can make the movements requested by its user *only* within the movements allowed in compliance of its regulations and instruction from dynamic traffic control systems.

2.7.2   Road Operator (A2)

The road operator is an agent of the local authority and/or central government, who instantiates and maintains the physical road structure, controls its traffic movements, and in some cases generates revenue from the road infrastructure.

2.7.3   Urban Authority (A3)

The urban authority may be the local government, or may be an agent of the local government, to provide administration to manage an urban area within the domain of the local government. They are classified as a separate actor because they are increasingly responsible for developing, and in some cases, operating, multi-modal transport systems, and, potentially, aspects of MaaS. In these paradigms they may provide additional control and management concerning the movement of automated vehicles within the urban domain.

2.7.4   Traffic Management Centre (A4)

The TMC is an agent of the local government, national road authority/ministry, or similar (according to the national implementation) who controls the flow of traffic through the road network. The movements of the automated vehicle are therefore within the direct control of the traffic lights, signals and other dynamic instructions of the TMC.

2.7.5   Automated Vehicle (A5)

The automated vehicle is an agent of the *vehicle user* who instructs it to make a journey from a defined source location to a defined destination location.

Note: in this taxonomy, the *vehicle user* of an automated vehicle is not identified as a separate actor because his/her objectives and actions are always identical to that of the automated vehicle within this paradigm. Clearly, within the J3016 levels, there are situations where the automated vehicle hands back control, or the vehicle user takes back control, but at that point it ceases to be an automated vehicle paradigm.

2.7.6   Connected Vehicle (A6)

Connected (C-ITS) vehicles are an agent of the *vehicle user,* which are under the control of a driver, but can communicate with an automated vehicle (and other connected vehicles), and can provide it with information/data that may help it with its driving task, and can receive similar information from the automated vehicle.

Connected vehicles will be an actor that shares the roads with the automated vehicle for the foreseeable future, except for the sub-category of automated vehicles that drive on segregated roads or separated road space.

2.7.7   Unconnected Vehicle (A7)

Unconnected vehicles are an agent of the *vehicle user,* which are under the control of a driver, but cannot communicate with an automated vehicle, and cannot provide it with information/data that may help it with its driving task, and cannot receive similar information from the automated vehicle.

Unconnected vehicles will be an actor that shares the roads with the automated vehicle for the foreseeable future, except for the sub-category of automated vehicles that drive on segregated roads or separated road space.

2.7.8   Emergency Responders (A8)

Emergency responders (ambulance, paramedics, firemen, police, etc.) are an agent of the local government that have priority access to the road network, and may have the ability to send instructions to an automated vehicle.

Emergency responders will be an actor that shares the roads with the automated vehicle for the foreseeable future.

2.7.9   Personal ITS (A9)

Personal ITS is a sub-category of the actor *pedestrian* (A21) that is able to receive, and potentially send, information/data via C-ITS-station from/to an automated or connected vehicle, probably (but not necessarily) using an equipped mobile phone.

Personal ITS will be an actor that shares the roads with the automated vehicle for the foreseeable future.

2.7.10   Map Provider (A10)

Map provider is an actor that provides map data, route data, and in many cases dynamic traffic data, to automated vehicles (and other actors and other road users).

Map providers may be government agencies, commercial organisations, or agents of vehicle manufacturers, or a combination thereof.

Map provider will be an actor that provides support to the automated vehicle for the foreseeable future.

2.7.11   Traffic Control Equipment (A11)

Traffic control equipment (communications directly >< vehicle) is an actor (road control equipment [traffic signals, etc]) that communicates directly >< with connected vehicles providing services (such as bus and emergency vehicle priority, green wave traffic light control, etc.). Automated vehicles, connected vehicles and vulnerable road users are actors that may use this actor's service.

2.7.12   Public Service Vehicle Communications (A12)

Public service vehicle (communications >< vehicle) is an actor that communicates with public services vehicles, primarily in the context of multi-modal journeys and MaaS. Automated vehicles and connected vehicles, are actors that may use this actor's service.

2.7.13   Travel Optimisation Service (A13)

The travel optimisation service / data centre actor is/may become appropriate in some instantiations of an automated driving system and in MaaS service provision. (see Chapter 11). (Other instantiations route this information via the map provider or MaaS service provider.) This actor accurately and frequently updates and makes information about regulations, events, situations, etc., available to connected and automated vehicles through its service provision. In may combine this with dynamic information from the TMC and static and dynamic information from public transport providers. It may provide this as data/information or may offer a travel optimisation service. In some instantiations *travel optimisation service* could become the automated driving system manager.

2.7.14   Cellular >< x Communications (A14)

The cellular >< x communications actor is included to recognise that 'connected' vehicles, including automated vehicles, may have other access to cellular connectivity (for example, for infotainment), operated by a third-party actor, that in the future may be used in ways we have not yet foreseen, and that could also introduce threats/r to connected and automated vehicles.

2.7.15   Vulnerable Road User >< Vehicle Communications (A15)

Vulnerable road user >< vehicle communications is an actor that is able to receive, and potentially send, information/data via C-ITS station from/to an automated or connected vehicle, using an equipped mobile phone or dedicated ITS station.

Examples of vulnerable road users are cyclists, micro-mobility scooter users and other micro-mobility device users, and mobility aids for those with physical disadvantages. The communications means may be a dedicated connected ITS station, but in many cases, will be an instantiation of *personal ITS (A9)* running a limited C-ITS station function as an 'app' on a phone.

Vulnerable road user >< vehicle communications will be an actor that shares the roads with the automated vehicle for the foreseeable future.

2.7.16 Independent Service Providers (A16)

Independent service providers (ISP) include motorists organisations/clubs, breakdown services, independent service stations, vehicle tuning services, car hire companies, fleet management service providers and sundry providers of other services that require access to data from the vehicle.

2.7.17 Insurers (A17)

Vehicle insurers offer use related insurance options, based on dynamic data from the vehicle. This may relate to distance driven, time of day, driver behaviour, etc.

2.7.18 Ride Hailing (A18)

Ride hailing services, currently such as Uber, Lyft, Wheely, Addison Lee, etc., that offer call on demand 'taxi' type services using a driver/chauffeur, but also including 'car sharing' services where the traveller has short-term on-demand car hire from fleets littered across the city. In the future, will include automated driverless vehicles providing an on-demand ride hailing service.

2.7.19 Micro-Mobility (A19)

Electric scooters, electric bicycles, public bike hire and other on-demand light transport services.

2.7.20 Public Transport/ Transit (A20)

Public transport services (buses, trams, trains, metro/underground systems, cable cars, lifts, funicular railways, etc.) operated by the local administration or operated by other authorities or commercial services offering transport on defined and declared routes, to a defined timetable and predetermined price. Novel PT services may also provide on-demand hailing transport with different pricing regimes.

2.7.21 Other Internet Communications (A21)

The other internet communications actor is included to recognise that 'connected' vehicles, including automated vehicles, may have other access to internet connectivity (e.g. for infotainment), operated by a third party actor, that in the future may be used in ways we have not yet foreseen, and that could also introduce threats or additional (as yet unforeseen) services, to connected and automated vehicles.

2.7.22 Pedestrians (A22)

A subset of pedestrians are instantiations of *personal ITS (A9)*. Apart from that subset, all pedestrians are unable to make contact with an automated vehicle other than by the operation of buttons on street furniture (except in the unfortunate event of making physical contact because the automated vehicle has collided with the pedestrian). In these circumstances they may be considered to be an instantiation of vulnerable road user. (But are not (A15) because there is no means of electronic communication available).

2.7.23   Drone & Kerbside Management (A23)

Local regulation and possibly control measures to manage rolling and ambulatory drones at a particular kerb or sidewalk; and management control of sidewalk (pavement) traffic and permitted levels of automation.

2.8   Other Functions

2.8.1   Regulation Data Access

Regulation data access is an important requirement that is made available by the *travel optimisation service (A13), map provider (A10), automated vehicle (A5)* (cameras) or a combination thereof, but it is not by itself an actor.

Vehicle >< vehicle communications

Vehicle >< infrastructure communications

Short/mid-range >< x communications

*Internet gateway*are communications means between ITS stations, but are not by themselves actors. No position is taken regarding which technologies are used.

A simplified communication taxonomy that embraces the complex reality of instantiation can be shown (without discriminating the communications means and technologies) and is shown below (Figure 2.7),

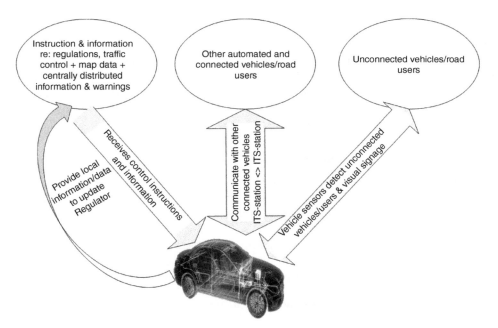

Figure 2.7 Automated vehicle system high-level communications taxonomy. Source: Bob Williams, Senior Consultant ,CSi (UK), Oxfordshire, UK.

In realisation, while the vehicle manufacturer's automated architecture diagrams only show the vehicle, they do embrace communication with other equipped vehicles, and they rely on the vehicle sensors to detect unconnected vehicles/road users/obstacles.

But they also rely on their sensors and map information/historic information records to identify control instructions and information, largely by expecting their sensors to 'read' road signs.

But as we have seen above, the automated vehicle is operating in a paradigm of control by regulations and instructions from TMCs. This is ignored in the paradigms of most vehicle manufacturers, and is a fundamental weakness. It means their design paradigm is fundamentally flawed. There will be more concerning this issue in later chapters.

3

The Current Reality

3.1 UNECE WP 29

Chapter 1 contained many of the (often extravagant) claims of automotive manufacturers. As described in Chapter 1, countries are also clamouring to take/claim leadership in introducing and testing automated vehicles and testing out MaaS concepts. So what is the current reality?

In September 2019, UNECE (United Nations Economic and Social Council Economic Commission for Europe) Inland Transport Committee, World Forum for Harmonisation of Vehicle Regulations published (2019) *(ECE/TRANS/WP.29/2019/34/Rev.1) "Revised Framework document on automated/autonomous vehicles".*

The primary purpose of the document is to provide guidance to WP.29 subsidiary Working Parties General Requirements (GRs) by identifying key principles for the safety and security of automated/autonomous vehicles of J3016 levels 3 and higher. Its safety vision states:

> … .for automated/autonomous vehicles to fulfil their potential in particular to improve road transport, then they must be placed on the market in a way that reassures road users of their safety. If automated/autonomous vehicles confuse users, disrupt road traffic, or otherwise perform poorly then they will fail. WP.29 seeks to avoid this outcome by creating the framework to helping to deliver safe and secure road vehicles in a consistent manner, and to promote collaboration and communication amongst those involved in their development and oversight.
>
> The level of safety to be ensured by automated/autonomous vehicles implies that 'an automated/autonomous vehicle shall not cause any non-tolerable risk', meaning that automated/autonomous vehicle systems, under their automated mode ([Operational Design Domain/OD]), shall not cause any traffic accidents resulting in injury or death that are reasonably foreseeable and preventable. Based on this principle, this framework sets out a series of vehicle safety topics to be taken into account to ensure safety.
>
> The report also goes on to require:
>
> System Safety: When in the automated mode, the automated/autonomous vehicle should be free of unreasonable safety risks to the driver and other road users and ensure compliance with road traffic regulations.
>
> Failsafe Response: The automated/autonomous vehicles should be able to detect its failures or when the conditions for the [Operational Design Domain/OD] are not met anymore. In such

a case the vehicle should be able to transition automatically (minimum risk manoeuvre) to a minimal risk condition.

Human Machine Interface (HMI)/Operator information:

Automated/autonomous vehicle should include driver engagement monitoring in cases where drivers could be involved (e.g. take over requests) in the driving task to assess driver awareness and readiness to perform the full driving task. The vehicle should request the driver to hand over the driving tasks in case that the driver needs to regain a proper control of the vehicle. In addition, automated vehicle should allow interaction with other road users (e.g. by means of external HMI on operational status of the vehicle, etc.)

Object Event Detection and Response (OEDR): The automated/autonomous vehicles shall be able to detect and respond to object/events that may be reasonably expected in the [Operational Design Domain/OD].

Operational Design Domain (ODD/OD)] (automated mode): For the assessment of the vehicle safety, the vehicle manufacturers should document the OD available on their vehicles and the functionality of the vehicle within the prescribed OD. The OD should describe the specific conditions under which the automated vehicle is intended to drive in the automated mode. The OD should include the following information at a minimum: roadway types; geographic area; speed range; environmental conditions (weather as well as day/night time); and other domain constraints.

Validation for System Safety: Vehicle manufacturers should demonstrate a robust design and validation process based on a systems-engineering approach with the goal of designing automated driving systems free of unreasonable safety risks and ensuring compliance with road traffic regulations and the principles listed in this document. Design and validation methods should include a hazard analysis and safety risk assessment for Automated Driving System (ADS), for the OEDR, but also for the overall vehicle design into which it is being integrated and when applicable, for the broader transportation ecosystem. Design and validation methods should demonstrate the behavioural competencies an Automated/autonomous vehicle would be expected to perform during a normal operation, the performance during crash avoidance situations and the performance of fall back strategies. Test approaches may include a combination of simulation, test track and on road testing.

Cybersecurity: The automated/autonomous vehicle should be protected against cyber-attacks in accordance with established best practices for cyber vehicle physical systems. Vehicles manufacturers shall demonstrate how they incorporated vehicle cybersecurity considerations into Automated Driving Systems, including all actions, changes, design choices, analyses and associated testing, and ensure that data is traceable within a robust document version control environment.

Software Updates: Vehicle manufacturers should ensure system updates occur as needed in a safe and secured way and provide for after-market repairs and modifications as needed.

Event data recorder (EDR) and Data Storage System for Automated Driving vehicles (DSSAD): The automated/autonomous vehicles should have the function that collects and records the necessary data related to the system status, occurrence of malfunctions, degradations or failures in a way that can be used to establish the cause of any crash and to identify the status of the automated/autonomous driving system and the status of the driver. The identification of differences between EDR and DSSAD to be determined.

(Source: UNECE WP29 GVRA Framework document on automated/autonomous vehicles ECE/TRANS/WP.29/2019/34/Rev.1 "Revised Framework document on automated/autonomous vehicles". © 2019, United Nations Economic Commission for Europe.)

The report additionally considers the requirements. In respect of automated vehicles regarding: Vehicle maintenance and inspection; Consumer Education and Training: Crashworthiness and Compatibility; Post-crash automated-vehicle (AV) behaviour.

3.2 Social Acceptance

In amongst the hype on the Mercedes' website are some perceptive statements such as The Verge (2019). *'Autonomous mobility will bring about major changes and involve many different parties. Psychological barriers must be overcome in the same way that social acceptance must be gained. This requires a sound legal basis across country borders that regulates autonomous traffic and covers questions of liability in the event of a collision'.*

3.3 SMMT

The UK Society of Motor Manufacturers and Traders (SMMT) Report 'Connected and Autonomous Vehicles 2019 Report/Winning the Global Race to Market' quoted above, went on to more conservatively state:

Level 5 automated vehicles should have the capability to be fully self-driving, unconditionally, and with no operating domain or geographic restrictions. Based on current technology roadmaps and real world applications, however, the consensus is that full and unconditional automation, i.e. Level 5, is unlikely to be introduced before 2035. One of the primary reasons stated by industry experts for this is the technology challenge involved in equipping AVs to tackle all possible unusual driving situations under all driving conditions and in all environments.

Instead, the road to Level 5 automated driving is likely to be reached gradually as more advanced driver assistance features come to market. This strategy, while incremental in its approach, is nonetheless expected to have a significant impact on the safety, convenience and cost aspects associated with current modes of transport. As this happens, disruption is likely to occur across traditional, ownership focused vehicles as well as shared mobility services such as taxis and shuttles. For example, it is estimated that there will be a 15% reduction in all collisions across major markets, including in Europe and North America, within a span of 10 years of Automated Emergency Braking (AEB) being mandated in Europe (expected between 2021 and 2025).

The report goes on to predict:

OUTLOOK TO 2040: BEYOND THE HORIZON

While the widespread rollout of Level 3 and 4 automation will likely create a significant impact on the UK economy by 2030, it is in the decade following 2030 that the most momentous changes will occur. The introduction of highly, and potentially fully, automated vehicles, the ubiquity of connected vehicles and the emergence of seamless MaaS business models will result in a complete overhaul of the way people commute, triggering a stronger impact on the overall economy.

One of the major influencing factors to consider while assessing the prospective impact on the UK economy is the overall growth in per-mile business models rather than per-car business models. The expansion of urban boundaries will make MaaS accessible to more people in the UK. All major OEMs are likely to have MaaS divisions focusing on revenue generation from new mobility modes and in-car data related services. Moreover, the market for premium vehicles in the UK is likely to grow further, with higher customisation leading to higher margins on vehicle sales.

Considering all of the important shifts in mobility, social and employment patterns, the overall impact on the UK economy due to CAV technologies could potentially be more than £145 billion by 2040. But, as referenced previously, the UK's exit from the EU must happen in a way that maintains the status quo as far as possible.

Similar to the previous decade, the economic benefits driven by CAV deployment are expected to accrue to end consumers, who will be able to better integrate their work and personal needs through seamless mobility modes and connected digital services.

Beyond the improvement in productivity, the adoption of CAVs is also likely to improve the overall convenience and quality of life for UK commuters as mobility will be more readily available to all and will provide a more stress-free commuting experience. This added convenience is likely to be the strongest driver to increase CAV adoption by consumers.

(Source: The UK Society of Motor Manufacturers and Traders (SMMT) issued a Report "Connected and Autonomous Vehicles 2019 Report / Winning the Global Race to Market". © 2019, SMMT.)

3.4 Other Observations

The UK newspaper, *The Guardian*, report commenting on the above report, also went on to state:

But the Guardian reporter also reported that 'However, Andy Palmer, the chief executive of Aston Martin, said he did not expect to see full autonomy in the next 20 years. He criticised measures announced to limit speed and said he believed manufacturers should go straight to level-four autonomy – where cars can essentially drive themselves and park safely should there be a problem – rather than bring in feature Fixes that remove some control from the driver.'

Palmer suggested the timetable for deployment of driverless cars had been exaggerated. 'You're going to see robotaxis in geofenced areas as early as 2021-22. I don't think you'll see commercial distribution of level-four vehicles until the mid-2020s. I don't think you'll see level five in my career. To drive up a mountain or a Delhi or London street – I think we're dreaming if we think it's going to be around the corner'. (Source: Gwyn Topham, Self-driving cars could provide £62bn boost to UK economy by 2030, April, © 2019, Guardian News & Media Limited.)

With each car manufacturer racing to develop its own self-driving solution, using their own in-house developed technologies, the behaviour of automated vehicles cars is becoming more and more fragmented. If this problem isn't tackled effectively, automatic vehicles will find it increasingly difficult to coexist safely, and pedestrians will not be able to anticipate the behaviour patterns of automated vehicles, and will therefore be at greater risk.

3.5 The European Commission

At the European level, the European Parliament and the European Commission have been enthusiastic to encourage automated vehicles, and particularly what they see to be the potential benefits associated with MaaS, to assist the policy aims of continued reduction in road deaths and injuries, and to reduce pollution, particularly in cities.

The European Commission *Strategy for Low-Emission Mobility* (European Commission, 2016) identifies the need for increasing the efficiency of the transport system by making the most of digital technologies, especially through cooperative intelligent transport systems (C-ITS) and successively, automated vehicles.

The European Commission, or more particularly, the member states of the EU, have been very successful in pursuing reduction in deaths and injuries caused by road accidents, halving the toll over a 20-year period. But the curve has flattened off over the last five years. In 2015, more than 26 000 people died and nearly 1.5 million people were injured on the roads of the European Union (European Commission, 2016). Worldwide, there were 1.25 million of road traffic deaths in 2013 (WHO 2015). As stated in the 10 goals for a competitive and resource efficient transport system (European Commission, 2011a), the EU aims to move close to zero fatalities by 2050. To achieve these goals, it is important to acknowledge that human error has been identified as a contributing factor in over 90% of all road accidents (Smith, 2013) and EC is desperately seeking new initiatives to reduce accidents. Automated vehicles, which eliminate human errors, are therefore an attractive proposition if they meet the safety and security objectives, such as those espoused by UNECE.

However, the reality to date, is less encouraging.

While automated vehicles show the promise to reduce accidents by removing human error, the coordinated automated road transport (C-ART) report (European Commission, Joint Research Centre, 2017) reported that *'it is not clear if they can compensate for crashes caused by inappropriate actions of other traffic participants (e.g., jaywalking pedestrians), vehicular defects (e.g., brakes failure), roadway factors (e.g., a pothole in the road), or environmental factors (e.g., fog) (Sivak and Schoettle, 2015). As a result, AVs may not be safer than an average driver and may increase total crashes in the mixed traffic period (Sivak and Schoettle, 2015). In (Schoettle and Sivak, 2015), it was found that AVs were involved in more crashes per million miles travelled than conventional vehicles although exposure was not sufficiently representative of the exposure of conventional vehicles. Other recent research found that AVs were involved in fewer crashes than conventional cars, especially for more severe crashes (Blanco et al., 2016 as cited in Townsend, 2016). In both studies, AVs that were involved in crashes were not at fault. The conservative behaviour of AVs could worsen the situation as they will tend to perform more cautiously compared to human-driven vehicles for safety and liability reasons. This circumstance may tempt human driven vehicles to adopt risky behaviours such as overtaking in dangerous situations or jumping in a platoon of AVs, thus introducing new risks'.*

Millard-Ball (2016) suggested that pedestrians will become less cautious and responsible around autonomous vehicles. Detailed analysis by Sivak and Schoettle (2015) concluded that autonomous vehicles may be no safer than an average driver and may increase total crashes when self- and human-driven vehicles mix.

The C-ART report (European Commission, Joint Research Centre, 2017) observes: 'The increasing amount of data which is generated, collected, processed and shared in our daily mobility holds enormous potential for the optimization of the transport system. The availability of this information in real time could considerably improve transport efficiency, sustainability, safety, mobility and comfort. Exchanging data between different actors in the transport system means supply and demand can be matched in real time, leading to a more efficient use of resources (European Commission, 2016). Intelligent transport systems (ITS) are advanced applications which without embodying intelligence as such aim to provide innovative services relating to different modes of transport and traffic management and enable various users to be better informed and make safer, more coordinated and "smarter" use of transport networks' (European Union, 2010). 'Applied effectively, ITS can save lives, time and money as well as reduce the impact of mobility on the environment' (Nowacki, 2011). As part of the digital single market strategy (European Commission, 2015), the European Commission aims to make more use of ITS solutions to achieve a more efficient management of the transport network for passengers and business.

3.6   Legislation

Legislation associated with traffic, driving connected and automated vehicles include:

- 1949 Geneva Convention on Road Traffic
- 1968 Vienna Convention on international road traffic
- Directive 2006/126/EC on driving license
- Directive 2003/59/EC on training and initial qualifications of professional drivers
- Directive 2009/103/EC on motor insurance
- Directive 85/374/EEC on product liability
- Directive 2007/46/EC on vehicle approval
- Directive 2014/45/EU on roadworthiness
- ITS Directive 2010/40/EU
- Directive 95/46/EC on data protection
- Directive 2002/58/EC on privacy in electronic communications
- Directive 2008/96/EC on infrastructure safety management
- UN Regulation No. 116 on anti-theft devices
- UN Regulation No. 79 for steering equipment
- UN Regulation No. 131 laying down the technical requirements for the approval of Advanced Emergency Braking Systems (AEBS) fitted on trucks and coaches
- Declaration of Amsterdam
- C-ITS communication (2016)

The *C-ITS Communication* from November 2016 (European Commission, 2016) highlights the relevance of cooperative, connected and automated vehicles for boosting the competitiveness of European industry, with a market potential worth dozens of billions of euro annually and hundreds of thousands new jobs created and for reducing energy consumption and emissions from transport.

> 'The development and deployment of automated and connected driving technologies should be supported by coherent transport research and innovation policies as well as appropriate regulatory framework conditions'.

3.7   Subsidiarity

'Subsidiarity' has a significant and complicating role in the implementation of automated vehicles and MaaS. Most vehicle regulations are developed by UNECE WP29. The EC, together with the member states, agree which of these regulations are to be adopted in Europe, but the actual implementation is a matter for 'subsidiarity', and is a national decision and implementation, and therefore liable to be implemented differently in different member states, especially in respect of issues such as traffic rules, rules on motor insurance and product liability, legislation for roadworthiness and maintenance, laws on ITS, data protection, privacy, infrastructure requirements.

3.8   Viewpoints

The Society of Motor Manufacturers observed in The Independent (2019): *'As this happens, disruption is likely to occur across traditional, ownership focused vehicles as well as shared mobility services such as taxis and shuttles. For example, it is estimated that there will be a 15% reduction in all collisions across major markets, including in Europe and North America, within a span of 10 years of AEB being mandated in Europe (expected between 2021 and 2025).'*

Whoever's hype and marketing position you tend to favour, it is clear that there will be a medium to long-term scenario where autonomous vehicles will have to co-exist and interact with large numbers of non-automated vehicles. Subsequent chapters will discuss issues where automated vehicles have problems with such interactions. But at this introductory level, what will be the impact be on safety caused by these issues? And just as importantly perhaps, what will be the impact on the public acceptance for automated vehicles when things do go wrong and there is an accident?

Elon Musk was once quoted to say that automated vehicles may make mistakes, but they will make less mistakes than human drivers, so automated systems are on balance safer. However, it seems unlikely that the public and regulators will make such a statistical assessment. Crashes involving automated vehicles will get blazoned on media headlines for many years, and may build up public resistance to the technology.

The most recent announcement regarding implementation (at the time of writing) is from German auto giant Volkswagen (VW) who unveiled the latest generation of its popular Golf compact passenger car with a host of new technologies, according to Automotive (Business Insider 2019/10 and Autonews, 2019).

Among the car's new features is vehicle-to-vehicle (V2V) and vehicle-to-infrastructure (V2I) communication technology, called Car2X. The Golf, the most popular passenger vehicle in Europe, will be the manufacturer's first model to boast the technology.

Car2X uses WLANp, a communication standard similar to Wi-Fi, to allow vehicles to communicate information between other vehicles and their environment. The technology will enable eligible Golf models to communicate road hazards or incidents with other Car2X-enabled vehicles or sensors within half a mile. For instance, cars can communicate instances of sudden braking ahead.

VW isn't the first to implement inter-vehicle communication, as automakers around the world are implementing V2V technology, though they're taking varying paths to do so. Beginning in 2015, Volvo partnered with regulators to test a limited number of Volvo 90 Series vehicles in Sweden and Norway equipped with cloud-based systems that enable the cars to communicate potential hazards.

In April 2019, Volvo announced that the systems would be available in its full 2020 vehicle line up in Europe. Meanwhile, in the US, automaker Cadillac has introduced V2V technology its models, but they can only communicate with other Cadillacs.

By deploying Car2X technology in the Golf, VW is likely attempting to build its case for making Car2X the preferred standard of vehicle-to-everything (V2X) communication. VW is a member of the Communication Network Vehicle Road Global Extension (CONVERGE) group, which is attempting to shape regulations and define an architecture for Car2X.

While VW isn't the sole automaker in CONVERGE – BMW and Opel are members too – Car2X still needs more buy-in from other automakers and cities to avoid fragmentation in standards for supporting vehicle communication infrastructure. This is especially key as regulators in the US and Europe are beginning to establish their preferred methods for V2X communications, though neither have come to a definitive conclusion – the European Commission has so far rejected legislation that would implement a Wi-Fi–based solution over cellular solutions, while US regulators have yet to mandate a standard.

VW's move to expand its V2V tech is a promising step, but it won't provide meaningful value for consumers or automakers until a significant number of cars on the road boast V2V capabilities. By introducing Car2X on a mass-market vehicle model, VW is likely aiming to make it appealing for connected infrastructure manufacturers to develop for as they chase a growing opportunity – the global V2X market is expected to reach $110 billion by 2026, up from $38 billion in 2018. If the company is able to secure a large developer network, it can more easily lure other automakers to use Car2X. (Source: George Paul, Volkswagen's upcoming Golf line-up will be able to talk to other cars and infrastructure, Oct 28, © 2019, Insider Inc.)

4

Automated Driving Paradigms

4.1 OECD

The OECD paper 'Automated and Autonomous Driving: Regulation under Uncertainty' (2015) identifies two major routes to automation. It describes the first route is described as 'something everywhere', which are vehicles that have some driver assistance (Level 1); these are already present today. It describes the second as 'everything somewhere', which is at the other end of the scale and refers to vehicles without a human driver and entails expanding the use of such vehicles to more contexts.

The OECD report concludes that 'These scenarios link to different business cases and use cases. High speed motorways may be promising for the early application of increasingly automated conventional cars and trucks (including platooning), urban areas are well suited for specialised passenger and delivery shuttles.' Within the context of these different scenarios there will be implications for other road users including cyclists, pedestrians and powered two wheelers (PTWs).

Although vehicle manufacturers seem to see a single paradigm for their automated vehicles (one where the vehicle is in sole control of its movements), the world is not that simple, and as we have seen in the previous chapter, the vehicle is just one actor in a controlled driving paradigm (see Chapter 2, Section 2.7), and a controlled actor at that. The OECD report highlights that the paradigm on intercity highways is very different than the paradigm within cities, and points out that automated vehicles will share the roads with nonequipped entities, in different ways, and will have to be operable within safe and feasible contexts.

It is clear that there is no one simple solution, but there will be an evolution over time, and in the meantime mixed traffic systems will have to operate. Much media attention, and a huge amount of government sponsored research, accompanied by hype for political purposes, is being given worldwide to automated and connected driving (Business Insider 2019/10 and Autonews, 2019).

4.2 Communications Evolution

While the researchers and standards development organisations and automotive engineers have been ploughing into the complex details of how to make such complex systems work coherently together, in an environment where the communication exchange must often have low latency – i.e. have very rapid

Automated Vehicles and MaaS: Removing the Barriers, First Edition. Bob Williams.
© 2021 John Wiley & Sons Ltd. Published 2021 by John Wiley & Sons Ltd.

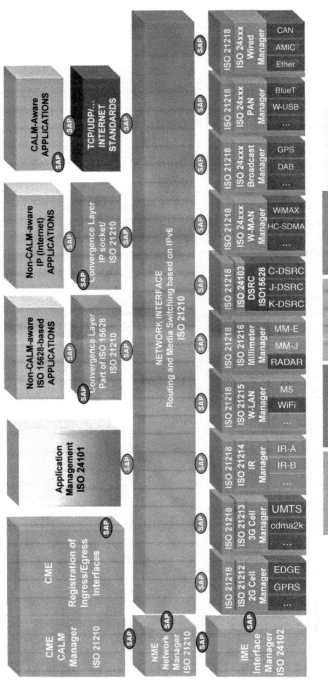

Figure 4.1 CALM 2001 architecture separating communications from applications (Evensen, Strasser, Williams). Source: "ISO 21217:2014 Intelligent transport systems - Communications access for land mobiles (CALM) – Architecture," March © 2014, ISO.

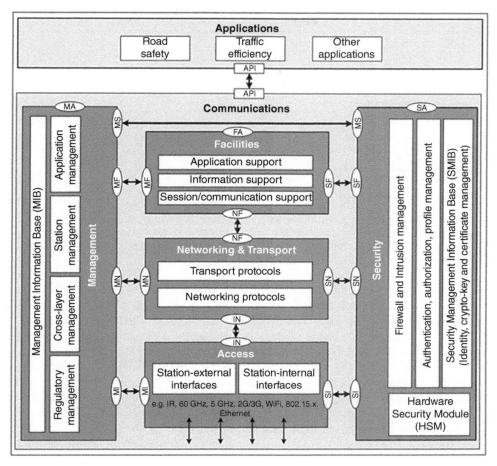

Figure 4.2 ITS station architecture (ISO 21217). Source: "ISO 21217:2014 Intelligent transport systems - Communications access for land mobiles (CALM) - Architecture," March © 2014, ISO.

response – telecommunications technologies have evolved rapidly. Back in the days of developing the CALM (communications architecture for land mobiles) architecture in 2000, telecommunications engineers were trying to handle the evolution from 2G cellular communications to 3G. Now 4G coverage is surpassing 3G, and 5G is in the process of being released. Significantly, modern wireless telecommunications have moved from 'circuit switched' to 'packet switched'. That is to say they have moved from being a telecommunications network in which two network nodes establish a dedicated communications channel (circuit) through the network before the nodes may communicate (The circuit functions as if the nodes were physically connected as with an electrical circuit) to a networking communication method where data is grouped into blocks called packets and routed through a network using a destination address contained within each packet, using internet protocols (IP).

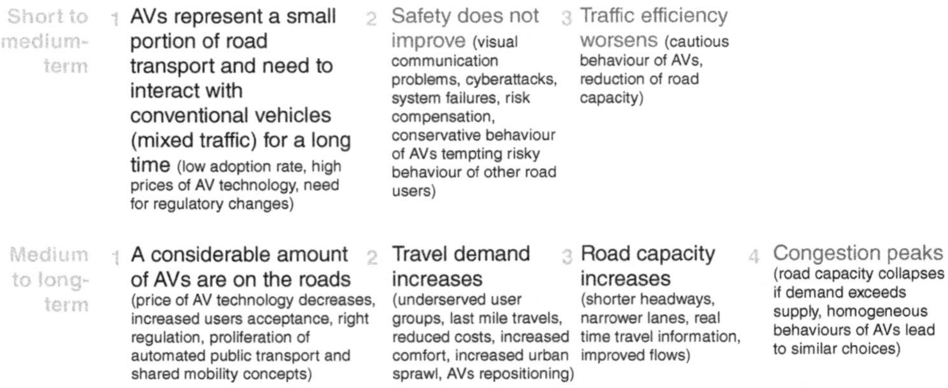

Short to medium-term	1 **AVs represent a small portion of road transport and need to interact with conventional vehicles (mixed traffic) for a long time** (low adoption rate, high prices of AV technology, need for regulatory changes)	2 **Safety does not improve** (visual communication problems, cyberattacks, system failures, risk compensation, conservative behaviour of AVs tempting risky behaviour of other road users)	3 **Traffic efficiency worsens** (cautious behaviour of AVs, reduction of road capacity)	
Medium to long-term	1 **A considerable amount of AVs are on the roads** (price of AV technology decreases, increased users acceptance, right regulation, proliferation of automated public transport and shared mobility concepts)	2 **Travel demand increases** (underserved user groups, last mile travels, reduced costs, increased comfort, increased urban sprawl, AVs repositioning)	3 **Road capacity increases** (shorter headways, narrower lanes, real time travel information, improved flows)	4 **Congestion peaks** (road capacity collapses if demand exceeds supply, homogeneous behaviours of AVs lead to similar choices)

Figure 4.3 C-ART Projection. Source: Alonso Raposo, M., Ciuffo, B., Makridis, M. and Thiel, C., The r-evolution of driving: from Connected Vehicles to Coordinated Automated Road Transport (C-ART), Part I: Framework for a safe & efficient Coordinated Automated Road Transport (C-ART) system, EUR 28575 EN,. Licensed under CC by 4.0.

The ISO 21217 ITS station architecture is technology agnostic. It is designed to function using more or less any wireless technology, and multiple technologies (for simultaneous support of multiple applications, simultaneously (so-called 'hybrid' communications)). The 21217 architecture will support applications designed to operate using multiple carrier technologies simultaneously, but that places demands of the application design, and none have, to date, been proposed. And it is designed in the expectation that new carrier media and additional carrier media will be introduced during the lifetime of the vehicle, so that an application can migrate from one carrier medium to another by software update at the application level, although some physical additions/changes may also need to be made (antennas, receivers, etc.).

Clearly, the future will be 'hybrid', i.e. using multiple communications means, to simultaneously support multiple service provision. But in order to realise low latency services, such as crash avoidance, ITS stations need to be communicating using the same communications medium, because the transactions, for obvious reasons, have to be very fast. But when do regulators and engineers fix on one technology for these day 1/1.5 low latency applications? In Europe, the C-Roads Platform, a group instigated by the European Commission at the end of the C-ITS Platform initiative has now committed (European Commission 2019/October) to the paradigm and instantiation based on:

> ……. *More specifically, first implementers from vehicle manufacturers, road authorities and road operators are addressing these issues on a large-scale with the provision of Day-1 C-ITS services (e.g., stationary vehicle warning, emergency electronic brake light, road works warning). The implementation is based on WiFi (ITS-G5) technology to improve road safety and reduce traffic congestion for European citizens as of today.*

The effectiveness of C-ITS is optimised by combining the ITS-G5 and standard cellular networks as part of a hybrid communication approach, which will reach more road users and larger portions of the European road infrastructure. This is why European C-ITS

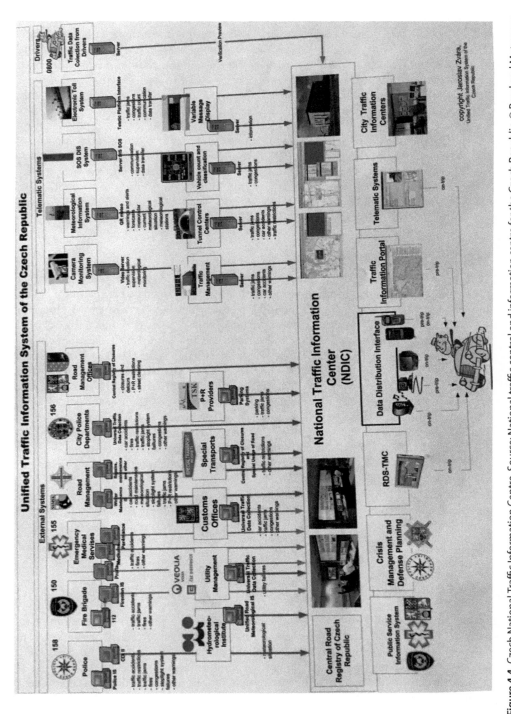

Figure 4.4 Czech National Traffic Information Centre. Source: National traffic control and information centre for the Czech Republic, © Road and Motorway Directorate of the Czech Republic.

*stakeholders such as road authorities, road operators, vehicle manufacturers, ICT indus-
try and the agriculture machinery and railway equipment sector are jointly committed
to C-ITS, that is based on available and proven interoperable harmonized specifications,
standards and technologies. There is large-scale deployment of C-ITS across Europe and
we will continue to deploy C-ITS using these proven wireless communication technologies, ensuring harmonization and interoperability of C-ITS services and applications.
This deployment approach is the baseline for wide European roll-out and the further
development of C-ITS services.*

> *....Supporters of the statement are committed to placing Europe at the forefront of
> development and deployment of CCAM relying on a strong regulatory framework.
> Deployment of C-ITS and ITS-G5 is progressing and therefore the first impor-
> tant step towards truly connected and automated driving has been taken!* (Source:
> From C-ITS deployment takes off, increasing road safety and decreasing conges-
> tion, 21 Oct. © 2019, CAR 2 CAR Communication Consortium.)

4.2.1   21-10-2019 Supporters

AEF, APCAP, ASFA, ASFINAG, Autopistas, AŽD Praha (Czech producer and supplier
of control and signaling technologies), Brisa, Brno City owned company for street
network maintenance and operation (BKOM), Brno City Public Transport Operator
(DPMB), CERTH-HIT, Car2Car Communication consortium, City of Vienna, City of
Kassel, City of Graz, CohdaWireless, Comune di Verona, Ministry of Transport of the
Czech Republic (MDCR), Czech Technical University in Prague, DAF, DARS, Dynniq,
GEVAS, Groupe Renault, HELLASTRON, ICCS, ITS mobility, INTENS Corporation,
KAPSCH, KTM, Magyor Közút, Land Salzburg, National railway infrastructure
manager in the Czech Republic (SŽDC), NXP, NeoGLS, O2 Czech Republic a.s.,
Ostrava City Public Transport Operator (DPO), State of Hesse, Pilsen City Public
Transport Operator (PMDP), RADOM Company, Road and Motorway Directorate of
the Czech Republic (ŘSD), Republic of Slovenia Ministry of Infrastructure, SANEF,
SCANIA, SEOPAN, ŠKODA Auto, Swarco, T-Mobile Czech Republic a.s., Transport
Infrastructure Ireland, Volkswagen Group, Volvo Group.

And the largest vehicle manufacturer in Europe, Volkswagen, is already equipping its
next release of its popular GOLF Model (GOLF 8), and has committed to equip all of its
vehicles from around 2022 (see Chapter 3). Other manufacturers are expected to follow
shortly.

4.3   Cooperative ITS

Cooperative ITS (C-ITS) otherwise called connected driving, is not a new phenomenon.
Back in 2000, ITS standards developers first faced the opportunities that wireless connectivity could bring, and the realisation that rapidly evolving communications technologies would change within the lifetime of a vehicle, and they set about creating an
architecture that could evolve though many generations of communications technology
through the lifetime of the vehicle, and one where more and more 'ITS services' became
possible.

This has evolved over time, and with input from many others, to the ISO 21217 ITS station architecture.

The past decade has shown significant progress in the provision of ITS services, now typically called advanced driver assistance systems (ADAS), (e.g. collision warning, lane keeping warning, blind spot warning, automatic emergency braking systems, etc.) and cooperative-ITS services (e.g. weather conditions warning, pot hole warning) await only agreement on the carrier technology for first-generation systems. The recent commitment, coordinated by the European Commission's 2015c-Roads initiative in Europe, the C-Roads Platform, a group instigated by the European Commission at the end of the C-ITS Platform (2016) initiative has now committed (European Commission 2019/October). Volkswagen's announcement that it will equip all of its new vehicles from 2022 with an ITS station capable of supporting G5, should break the log-jam.

The European Union has particularly identified first C-ITS, and latterly automated driving and MaaS, as a means to significantly reduce road deaths and injuries. The EU have created several initiatives to encourage and implement these technologies. Amongst others:

- G7 declaration on Automated and Connected Driving (European Commission 2015a).
- Round Table on Connected and Automated Driving (European Commission 2015a)
- The C-ITS Platform (2014-6) (2016)
- Gear 2030 initiative (European Commission 2016b, GEAR 2030, 2016)
- Declaration of Amsterdam (European Union 2016)

The European Commission generally believe that increased level of connectivity and autonomy of road vehicles could lead in the future to a complete transformation of road transport. The European Commission provides significant research funding to try to stimulate these advances.

4.4 The C-ITS Platform

The C-ITS Platform (2016) recognised the SAE J3016 levels of automation, and formed the recommendation that next steps should be focussed on getting on with actually delivering day 1/1.5 services rather than eulogising about the long-term potential.

The C-ITS platform agreed on a list of 'Day 1 services', which, because of their expected societal benefits and the maturity of technology, are expected to and should be available in the short term (personal benefits, users' willingness to pay, business cases and market-driven deployment strategies were not taken into account at this stage):

List of Day1 Services

- Hazardous location notifications:
- Slow or stationary vehicle(s) and traffic ahead warning
- Road works warning
- Weather conditions
- Emergency brake light
- Emergency vehicle approaching

- Other hazardous notifications
- Signage applications:
  - In-vehicle signage
  - In-vehicle speed limits
- Signal violation / intersection safety
- Traffic signal priority request by designated vehicles
- Green Light Optimal Speed Advisory (GLOSA)
- Probe vehicle data
- Shockwave Damping (falls under ETSI Category 'local hazard warning')

Furthermore, the C-ITS platform also agreed on a list of 'Day 1.5 services', considered as mature and highly desired by the market, though, for which specifications or standards might not be completely ready.

List of Day 1.5 Services

- Information on fuelling and charging stations for alternative fuel vehicles
- Vulnerable road user protection
- On-street parking management & information
- Off-street parking information
- Park & Ride information
- Connected and cooperative navigation into and out of the city (first and last mile, parking, route advice, coordinated traffic lights)
- Traffic information and smart routing

These, the C-ITS Platform, opined, were early deliverable 'connected vehicle' services that would provide the gateway to more ambitious services. But they are clearly J3016 Level 1 and Level 2 services, and far from the J3016 Level 5 'fully automated' driving paradigm.

When describing and comparing viable paradigms for automated driving and MaaS, we have to identify what we mean by 'viable' and limit ourselves to such options. In its report 'The future of Mobility' (2019), the UK Department for Transport identified key criteria to be required/assessed to identify a permissible and allowable future mobility paradigm:

In facilitating innovation in urban mobility for freight, passengers and services, the government's approach will be underpinned as far as possible by the following principles:

1. New modes of transport and new mobility services must be safe and secure by design.
2. The benefits of innovation in mobility must be available to all parts of the UK and all segments of society.
3. Walking, cycling and active travel must remain the best options for short urban journeys.
4. Mass transit must remain fundamental to an efficient transport system.
5. New mobility services must lead the transition to zero emissions.
6. Mobility innovation must help to reduce congestion through more efficient use of limited road space, for example, through sharing rides, increasing occupancy or consolidating freight.
7. The marketplace for mobility must be open to stimulate innovation and give the best deal to consumers.

8. New mobility services must be designed to operate as part of an integrated transport system combining public, private and multiple modes for transport users.
9. Data from new mobility services must be shared where appropriate to improve choice and the operation of the transport system.

This is, of course, a politically affected list designed to achieve policies of the UK government at its date of issue. However, it has a modicum of common sense, and its first 'Principal' will surely be the key principal for any regulator. 'New modes of transport and new mobility services must be safe and secure by design.'

This will be the approach of any regulator. But note it is not 'must be safer than the current paradigm' but '… must be safe and secure by design'.

4.5 Holistic Approach

As elements of the transport system become increasingly interconnected, a holistic approach is needed in order to intermesh and enable the complex interactions among different players such as vehicles, drivers, infrastructures, policies, citizens, energy, economy and environment. Vehicle manufacturers ploughing lone furrows to introduce automated vehicles and MaaS, using different paradigms, and defective architectures, will delay or prevent the introduction of these technologies. Teslas only communicating with Teslas, and Fords communicating only with Fords, is a hopelessly defective paradigm. Yet this is how auto manufacturers are developing their 'extended vehicle' cloud instantiations.

AVs by themselves will not necessarily be smarter than conventional vehicles driven by humans. A human driver, given 12–36 hours of driving lessons, and a bit of practice on the road, and a written test, sits a driving test, and a high proportion are judged fit to control a vehicle without further supervision on the highway. Google, on the other hand has been driving in full automation mode for more than 2 million miles, mostly in urban environments, and, at the time of writing, is still working to get it right.

We can assume that AVs will follow more rational rules than humans. But without coordination with other vehicles and the infrastructure, AVs will try to find to find their individual optimal solution in an analytical way. The existing heterogeneity of vehicle behaviour, which is considered one of the main causes of traffic instability Ngoduy (2013a,2013b) is unlikely to be reduced, because the 'go-it-alone', competitive strategies of vehicle manufacturers will continue to be keep vehicle technologies differentiated, and each vehicle manufacturer will implement (and keep strictly confidential) its own driving logic.

In theory, the reaction time of an automated vehicle can be significantly less than that of a normal driver, leading to a significant increase in road capacity Kesting and Treiber (2008). However, liability issues on the responsibility of automated vehicles are likely to force vehicle manufacturers to design their vehicles to be fairly conservative, a serious problem when mixed vehicles (automated and conventional) will be on the roads together.

Technology has advanced enough that I think that we can more or less be confident that by 2022, automated shuttles will be deployed on segregated roadways, often mixed with pedestrians, and will function as low-speed shuttles from bus or parking stations to shopping malls, around resorts, animal, and theme parks, etc. They may be small 'pod' shuttles, or more commonly small buses. But they will rarely travel on mixed road environments as part of a mixed flow of traffic, except perhaps where the speed limit for all vehicles is very low. It will probably be enough to wipe the egg off politicians' faces when widespread use of fully automated vehicles fails to appear on our highways in the early 2020s; but some over-optimistic automotive chiefs are likely to forced to consider a career change.

It has been proposed that some roads could be allocated exclusively to AVs, or special roads (with very high standards of road marking), built for them. Other ideas being discussed are that some HOV (high occupancy vehicle) lanes could be reassigned for exclusive use of AVs, that motorways could have one lane allocated to automated and 'connected' vehicles, etc.

If you believe these possibilities to be unrealistic, consider the introduction of autoroutes/motorways to adapt to the spread of ownership of cars, or the laying of by-passes, ring roads, rocades, etc., to lessen the flow of traffic through city centres. All of which happened extensively in the second half of the last century.

4.6 It Won't Happen Quickly

But even as this happens, it won't happen quickly. While AV manufacturers are claiming to get their vehicles into production by 2020/2022 – an optimistic target date that they won't make – it ignores the fact that changes to road allocations, as well as being the subject of potentially long political debate, then have to pass through the planning applications system (different in every country, but ubiquitously bureaucratic), and then the roadwork has to be undertaken to enable this . . . . In practice just changing a speed limit and putting up the new signs can take four to five years. Even more so when it comes to laying new roads, or widening motorways to accommodate lanes for automated vehicles. (The M25 around London took a quarter of a century to complete). So, the timeframe for such radical changes would likely put an implementation back to at best around 2025–2030 but more probably beyond. . . . . Maybe a lot further beyond.

But even if these things come to pass, as the OECD report pointed out, one of the key challenges along the path to full automation will be how, in the interim, automated and semi-automated vehicles co-exist with nonequipped vehicles. Tesla probably has the most experience here, and their system seems workable most of the time, and is undoubtedly very clever, advanced, leading edge technology – but unfortunately it does not work all of the time, and they still crash into stationary vehicles, have trouble in high-contrast light situations, and sometimes don't recognise/differentiate humans, animals, or plastic bags wafting in the breeze, etc., in their path, that mean that they can only be allowed to operate 'automatically' (Tesla calls it 'autopilot') under the watchful supervision of the driver, which kind of defeats the objective.

The stark reality is that this interim period, even optimistically assuming that everyone will use automated vehicles one day, will last at the very least 15 or more years, probably more than 25 years, possibly 50 or more years, because cars last 15–20 years, trucks and buses even longer, and that is how long the replacement cycle will last. And the true reality is that it is unlikely that automated driving will take over entirely. European drivers can't be persuaded to hand over the simple changing of gears to automation, and most cars sold in Europe are still manual gear shift......... I have to be convinced by some new, as yet unknown, factor that will persuade all European drivers to all hand over complete control to an automated vehicle. So, the probable realistic paradigm is that automated vehicles and non-automated vehicles will have to co-exist (safely) for the foreseeable future.

Safety evaluations and predictions are usually based on assumptions of a fully equipped fleet of comparable vehicles and it seems that very little research has been conducted on the safety aspects safety impacts during this transitional phase. This is not surprising because the paradigms that (apart from some automated shuttle-bus developers) those automotive manufacturers, or technology providers who want to be driverless vehicle system providers, who are designing and building these AVs, are designing to a paradigm that they are in control of their environment, when the reality, as we have seen above, is that they are not. They are working in a paradigm controlled by others.

With little cooperation between AV manufacturers, how AVs react to the close proximity to each other and how they react to 'incidents' and surrounding conditions, is crucial.

In the paradigm of governments, researchers and standardisers, these vehicles are in communication with each other in real time via an 'ITS station' in one vehicle communicating with an 'ITS station' in another, so there should be no problems, (though further work is needed).

However, the 'go-it-alone', 'autonomous' direction of most AV developers have avoided much of this cooperative approach, and in so doing are introducing considerable uncertainty and risk, that directly compromises the requirement that these new systems '..... must be safe and secure by design', thus threatening their acceptability.

4.7 Implications of Fully Automated Vehicles

Driverless vehicles are often described as a 'new mode of transport' and will bring disruptive change to travel patterns and changing mobility culture. Research from several US research projects on the implications of fully automated vehicles for vehicle ownership and use found that they may lead to a reduction in vehicle ownership of over 44% due to increased vehicle sharing. Moreover, the same research found that this could also lead to a large increase of 75% in individual vehicle usage Schoettle and Sivak (2015).

But how much of this research is truly objective, and how much are the optimistic hopes of advocates?

A lesson learned by this author a couple of years ago when we changed one of our cars. The new car was of course crammed full of ITS services, as you might imagine, and was partly selected for that reason. Trying to explain the features of our new car to my (then) eight-year-old daughter, and relate it to what Daddy did for a living, she responded with what features she would like to see in her first new car when she gets one when she grew up.

At that point I explained that in 10 or 15 years, she might not want to own her own vehicle, and I embarked on a high-level, simplified, explanation of MaaS. In response, she gave me one of those withering looks that only a daughter can give to her father when it was obvious that he just didn't get it.

'But Daddy', she said, 'We have that service already, it's called a taxi, and you don't like using them because they never show up at the appointed time, the driver talks on the phone when driving, they are too expensive, often not very clean, and the driver doesn't speak English properly. I want my own car, with my own space, my own things in the car which are mine and make my journey comfortable. I don't want to be in someone else's smelly car that smells bad.'

Forecasters, have your dreams. But beware. 'Out of the mouths of babes and suckling's …' oft comes a dose of realism. MaaS has its place, but it will not take over entirely.

Also, and just as important as how well AVs interact with their environment, is how other road users will interact with AVs.

This thought raises another challenging question. How will vehicles with speed management systems interact with unequipped vehicles? Will the unequipped vehicles travel faster and continually overtake so that the speed management system gets finally switched off by a dissatisfied driver/user? Or if he cannot do that, or it is an automated vehicle, will its journey continually be pushed back by nonequipped vehicles pushing in front of it, and will the accident risk increase sufficiently because of imprudent overtakes by frustrated unequipped drivers, that the automated vehicle will slow down further and further because of the additional risk. How is that going to work out??

Within this overall requirement, if we want to understand the possible road transport system responses with the introduction of AVs, we then need to understand the fundamentals that will increase or decrease travel demand and road capacity. The effects will differ according to the level of penetration of automated driving technologies, but need to be considered as part of the process of providing 'viable' paradigms.

Travel demand could move in in two directions, i.e. increasing or decreasing, with AVs.

If AVs are, as claimed, to lead to more efficient use of vehicles, it would seem logical to believe that there will be less congestion, therefore less time spent on journeys and a better cleaner environment, and reducing the cost per mile travelled, so more cost-efficient, which explains the appeal of AVs to politicians.

But the result of making road travel cheaper, more comfortable, more efficient and accessible to new user groups, may be that travel demand (provided by MaaS) could potentially increase overall travel demand. But, as a consequence, use of public transport could decline, affecting their viability, leading to a reduction of public transport options available, leading to yet further increase in road miles driven.

Increased demand is both because existing latent demand from these new (underserved) groups of users can now be satiated, and the creation of new demand resulting from capacity improvements enabled by AVs (similar to the effect that a new bypass or motorway increases total traffic movements). More attractive travel conditions, and the fact you can work during the journey, will encourage longer commutes, thus further increasing road miles driven.

Against that, it is claimed by advocates, if there is an increase in ride sharing, this could reduce vehicle ownership and travel demand. But why would there be an increase in ride sharing just because the vehicle is automated? The growth of ride sharing (i.e. two or more people agreeing to share a single vehicle to make the same commute) is based on other much stronger factors such as sharing the cost, or use of HoV lanes, costs and availability of parking in the city, or social political pressure or trendiness, than whether or not the vehicle is automated. So to this authors view, such logic is faulty. Ride sharing will decrease road miles driven. That is faultless and obvious logic, equally with driven and automated vehicles. It is a fact, but it has nothing whatsoever to do with the uptake of automated vehicles.

As to the effects on the roadway infrastructure, road capacity could theoretically increase with a higher penetration rate of AVs and connected vehicles because these vehicles can adopt shorter headways between vehicles (e.g. platooning), and provide a better traffic distribution using real time traffic information. But in the mixed traffic of AVs and conventional cars, short headways may also encourage drivers of conventional vehicles to adopt unsafe practices. Therefore, as with adaptive cruise control, these opportunities may not be taken up by system designers, in order to better manage their risks/liabilities. They will play it safe in their designs. The adverse consequences are too costly not to do so.

There has been much publicity that MaaS using automated vehicles will reduce the demand for in-city parking. And combined with policy shifts to encourage the use of public transport in cities, that seems a likely outcome. However, it will come at the expense of *increasing*, not decreasing, traffic in towns.

This is because a commuter travels X miles to commute from home to work, where he leaves his vehicle in the car park all day, then travels X miles back home at night. If he uses an automated vehicle in a MaaS environment, the vehicle has to first travel Y miles to get from its depot or last drop-off to our commuter. Meaning that in the drive your own car paradigm, 2X miles are driven each day. In the MaaS paradigm, this increases to 2X + Y, and Y can be a significant addition.

The next effect to consider in a little greater depth is the psychological effect on the commuter of using an AV. The automotive manufacturers claim (and for once can't be criticised for the claim), that if the driving process is completely automatic, the users of the vehicle are freed up to do other things. Thus (s)he could perform part of his/her daily work task in the vehicle – allowing to set off later and return earlier, or extend the commute to live somewhere more desirable or, in the case of large cities, more affordable. With the consequence of increased road miles driven.

According to some authors Schoettle and Sivak (2015) AVs operating in a MaaS paradigm could induce an increased travel per vehicle of up to 75%, even if vehicle ownership could be reduced up to a 43%. MaaS AVs repositioning that result from AVs travelling empty to pick up passengers could increase travel distance by 11% compared to privately owned vehicles Fagnant and Kockelman (2014).

Project C-ART (https://www.darpa.mil/news) summarised that '... some authors have estimated that the reduced cost of driver's time in AVs could result in an increase in light duty vehicle travel between 30% and 160% (MacKenzie et al. (2014) as cited in LaMondia et al. 2016), while others indicate changes in vehicle kilometres travelled (VKT) ranging from a 4% increase for low-level automation to around 60% increase for high level automation (Wadud et al. 2016).'

OECD/ITF 2015 estimates suggest travel increase between 30% and 90% with mixed-fleets of shared AVs and traditional private cars, potentially also including a rise in the number of vehicles.

This huge increase is a recognition that there is a very significant underserved potential. Those under the age to drive, those too old to drive (a hugely growing proportion of the population, that could be over a third of the population within 50 years), the disabled, and, not to be forgotten, the large number of people who just do not have a driving licence.

With ever busier airports providing a less pleasant travelling experience, and social pressures to reduce air travel (albeit a current social pressure that may change), alternative travel means provide an opportunity for a better travel experience. Comfortable AVs totally redesigned from current vehicles, and being exhibited at motor shows as concept vehicles by automotive manufacturers, offer the possibility to lounge, relax, watch films, even lie flat and sleep (mimicking airline business class modules) may make AVs a viable alternative for land journeys, thus further increasing load on the road network. La Mondia et al. (2016) in a paper to the Transport Research Board in the United States, postulated that, a significant rise in additional long distance trips could be expected with AVs, for trip distances below 500 mi (this equates to 800 km).

The so-called but misnomered 'rideshare' taxi-hailing services, such as Uber, (misnomered because the hirer does not normally share his/her trip with other unknown passengers), together with home-delivery services, such as Amazon, have been praised by many sources as resulting in less use of personal vehicle kilometres driven, and a significant reduction of coming miles driven to large supermarkets. But the net result is that they increase, not decrease, congestion, and have been proven to increase congestion, for example, in London.

The Transport for London (TfL) performance report on the roads it controls, which covers the third quarter of 2015/16 (n.d.), records a 'significant slowdown in the rate of traffic growth in London' but also 'a significant deterioration in London-wide traffic speeds', quantified as a 7.7% reduction compared with the same period of the previous year.

Increases in the numbers of private hire vehicles (PHVs) and delivery vans have contributed too. A linked TfL reported stated 'that the number of PHV (Private Hire Vehicle) drivers has increased from 59,000 in 2009/10 to more than 95,000, that the number of PHVs circulating in the Central London congestion charge zone has increased by over 50% in the last two years, and that one in ten vehicles entering the zone is now a PHV'. A recent study published found that the growing e-commerce market is also increasing gridlock, with a numbers of light goods delivery vehicles in the centre of London soaring.

We have then to postulate, during a period where AVs are very much the minority, the effect on mixed traffic conditions. As argued above, AVs may reduce roads capacity in the near term, e.g. by maintaining large headways and, as much for liability reasons, and to meet the requirements of governments who conditionally allow AVs on the road with one of the conditions being that 'New modes of transport and new mobility services must be safe and secure by design', by reacting tentatively after yielding, leaving generous margins to allow for the weaknesses of human drivers, or stopping, thus reducing the available space for other vehicles. We can predict this reasonably reliably, because we have ample evidence that rain, sleet, snow and fog reduce the capacity of the network because of (understandable) increase in caution, and these are behaviourally similar consequences.

Once in the flow of a continuous traffic stream, we have then to consider the different ways that human drivers and AVs react to congested traffic streams. The first part of that analysis is probably not too difficult. Most traffic modelling, by too many sources to quote, conclude that the overall movement of traffic is generally best achieved, and destination reached more quickly, by maintaining the same lane. So, it is not too speculative to predict that AVs will tend towards this model. However, it is also known not just from research studies, but every driver's own experience, that regardless of this, a significant proportion of manual drivers 'weave' lanes in the belief that this will give them relative advantage.

But what is not known, and this author has seen little but speculation, is the effect that weaving traffic will have on AVs? Weaving drivers tend to take significant risk, relying on the following (intruded) vehicle in the new lane to take whatever action is needed to not collide with the intruding vehicle. We know that this creates disruptive shockwave ripples and that a relatively few of these events cause significant slowdown of the traffic flow. However, we know from accident reports that in these common situations, both intruder and intruded parties take more risk than they usually consider safe,

Following the principal of 'New modes of transport and new mobility services must be safe and secure by design', we can only guess that AVs will react on the safe side taking less risk than the manually driven intruded vehicle. We can also guess, and it can be no more than a guess at this stage, that this caution will cause stronger shockwave ripples that further exacerbate the congestion.

As the proportion of AVs in the traffic stream rise, this trend is expected to exacerbate the situation (congestion) even further, and will not probably improve the situation until AVs become the majority. (But this is very speculative).

In the UK (Sabur 2017) has estimated a rise in delays on motorways and major roads during peak periods by 0.9% when a quarter of vehicles are automated. Project C-ART (https://www.darpa.mil/news) predicts that congestion levels will not drop significantly until AVs make up between 50 and 75% of the vehicles fleet. But, as stated above, at this stage, both of these predictions must be considered very speculative. And we have seen also in the arguments above, that automated vehicles may never form 75% or more of the vehicles on the road.

ETSC (European Transport Safety Council), in a recent briefing, 'Prioritising the Safety Potential of Automated Driving in Europe' (2016) predicted 'In general, it is expected that the first vehicles with full or advanced automation, which will only operate within limited areas, will become commercially available in the early 2020s. .... Fully automated vehicles that operate on public roads among other traffic are unlikely to be on the market before the 2030s .'

Project C-ART (https://www.darpa.mil/news) concluded that 'these factors, linked to the high average ages of vehicles on the road, indicate that there will probably be decades during which conventional vehicles and AVs would need to interact. This period could even last indefinitely as some people may want to drive only conventional vehicles (Schoettle and Sivak 2014 as cited in Sivak and Schoettle 2015). In this mixed traffic period, conventional vehicles will interact with AVs of varying levels of automation (with probably more amount of those with lower levels of automation).'

Research (Sivak et al.) has also been undertaken which, if true, dispels the high expectations of automated driving as a tool to reach the road safety goal of zero deaths. It raises other influencing factors that an automated vehicle will struggle to deal with. These researches also argue that automated vehicles will find it hard to perform perfectly, for example, under all weather conditions or in cases of crashes being caused by other traffic participants, and, for example a pedestrian stepping out unexpectedly in front of the vehicle, at close range. And these factors may serve to increase, rather than decrease road deaths and injuries, and, in any event, will to some extent offset gains made,

On the other hand, it is well researched and agreed that most crashes involve some element of human error. If greater autonomous operation reduces or eliminates these errors, then benefits for road safety may be substantial. ETSC endorses the 'safe system' approach meaning that 'human beings are fallible, and their errors must be anticipated and the risk of serious consequences from these errors minimised.'

And ETSC raises the societal issue that 'The responsibility for reducing fatalities and serious injuries is therefore not solely placed on the road users but shared with e.g. vehicle producers and infrastructure managers.' Thus, automated driving can be welcomed as a way of further sharing the responsibility to vehicle manufacturers and infrastructure managers in the future.

At present there are many different circumstances that can lead to a driver's inappropriate situation assessment, inattention or distraction. These have been calculated by various researchers as contributing to as much as 10–30% of road deaths.

Increased levels of vehicle automation could contribute to eliminating or easing conflict situations. It is expected that it could make a contribution by reducing visual error, single-vehicle crashes, and crashes at intersections. Automation could be expected to reduce some high-speed collisions on the motorways due to the fast reaction times. It could also address fatigue related crashes, although driver operator sleepiness may be enhanced due to boredom and to disengagement from vehicle control.

However, the OECD report argues that the real safety test for autonomous cars will be how well they can replicate the crash-free performance of human drivers. As drivers gain experience, they master not only the physical control of the vehicle, but many develop a road alertness that identifies irregular behaviour of a nearby vehicle. It may have broken no rules, but it is just noted as different by the alert driver. Approximately 99.99% of the time nothing occurs, but just every now and then that intuition enables the driver to avoid an incident. It is hard to capture 'intuition' in an algorithm, and it is yet to be seen if AVs can as effectively learn such skills.

Whilst the discussion above may seem somewhat unstructured, the conclusions that can be drawn are that the introduction of AVs will have a general benefit for the traveller, but the outcome on the road network cannot be seen as simple and there will be competing benefits and disadvantages that are probably too complex to predict anything other than that 'there will be competing benefits and disadvantages'. Simplistic statements, like 'AVs and MaaS will reduce pollution and congestion' should be avoided. What is reasonably sure is that AVs and MaaS will bring change, probably disruptive, and probably simultaneously beneficial and adverse, to the way we think about the travel experience, the way we travel, and will have impact on the road infrastructure; and that these effects will be different at different rates of AV penetration.

The EC JRC Project C-ART (https://www.darpa.mil/news) made the following projection: Project C-ART explained its projection for expected impacts of AVs in the short to medium term (2020–2030) as:

> AVs would represent a small portion of the overall vehicle fleet and they would need to interact with conventional human-driven vehicles for a long period of time

In the short-term, it is likely that the proportion of AVs on the roads will not be significant. One reason for this will be the high prices of AV technologies, since even if they will likely become cheaper with mass production, they will probably be relatively expensive as they need to meet high manufacturing, installation, repair, testing and maintenance standards (Litman 2016). In addition, subscription fees for special services (e.g. mapping) may be required.

C-ART also indicated that early research (there are no automated vehicles on the roads yet apart from a few R&D vehicles, so all research to date must be considered 'early research') indicated that there was likely to be a low level of trust on the safety of these vehicles, and a reluctance to use them, preferring the trusted solution of a human driver. Schoettle and Sivak (2014) concluded that opinion was widespread that that AVs would not drive as well as human drivers and that users would feel uncomfortable travelling in an AV, and this has been backed up by other research. Users' acceptance towards AVs will of course grow as they become more familiar on the roads, but initial fears will inhibit early take up and affect early business models.

Connected C-ITS vehicles face fears about data anonymity but do not face the same reluctance to trust the technology.

Of course, automated and connected vehicles bring new requirements and challenges to the regulators in respect of traffic law, liability, security, access to data, protection of personal data, particularly for Level 4 and Level 5 AVs.

And C-ART concludes that 'During these mixed traffic conditions, safety would not improve. During the mixed traffic period, communication problems are expected between AVs and human driven vehicles. Conventional vehicle drivers' expectations about likely actions of surrounding AVs will be affected, mainly by the lack of eye contact feedback (Sivak and Schoettle 2015), and by reactions that are unexpected to human drivers. It is well known that most of the accidents involving Google cars have occurred due to the incapability of human drivers to anticipate reactions such as the sudden stop of the leading AV due to leaves or shopping bags fluttering on the road. Time will be needed for human drivers to understand how different AVs behave and therefore it is hard to believe that AVs will contribute to safety from the first moment they appear on the road.'

Assuming the C-ART predictions to be in the right ballpark, we have then to consider the effect in the short/medium turn, noting that in this paradigm, 'short to medium' cover not Ford's and Tesla's claims to have fully automated vehicles on the roads by 2021, but covers at least the whole of the 2020s and maybe the 2030s, i.e. the next 20 years, maybe longer. Maybe much longer.

While this book has criticised automotive companies for their naïve understanding of the automated driving paradigm, now that governments and city authorities realise the opportunities and potential benefits that MaaS can bring, if they are the controllers of the road network within which automated driving and MaaS will occur, they have now to step up to the plate and deliver the information that is required for automated driving and MaaS (there is more of this discussion in Chapters 6, 7, and 9).

Keeping the discussion at a high level, we talk of V2V (vehicle to vehicle) and V2I (vehicle to infrastructure) communications, and much has been said about potential V2V services. But we now need to pay more attention to the potential roles that a local authority, central government or commercial service operator, or a combination of the three, will provide.

The Czech Republic are probably furthest ahead on these considerations and already have a 'National Traffic Information Centre' (operating in Ostrava).

While this centre predates C-ITS implementation, it shows the collation of data in digitised form from many sources, and the provision of that data to the vehicle through multiple means, some pre-trip, some on-trip/dynamic. It is of course readily extensible to accommodate C-ITS data provision.

In other countries, political decisions need to be made as to whether the provision of data and other C-ITS service provision to vehicles from the infrastructure should be a local authority service, a regional service or a national service, or a commercial service, or a combination of all four.

The operator of the infrastructure C-ITS station will then have to identify what ITS services it will support and how. While lists of day 1 and day 1.5 services have been agreed by the ITS platform report, and generally described, and the HARTs ITS architecture characterises likely inputs and outputs, no detailed system analysis of any of these services has been circulated, and no standard service definition issued. These steps need to happen very quickly. And these services need to be defined clearly, including data tree definition and unambiguous data definition, probably in ASN.1.

The C-ART project (https://www.darpa.mil/news) takes things further, advocating that a central facility should actively manage and control vehicle movements in its domain. C-ART foresees three phases: First of all, a shift from conventional vehicles to connected vehicles. Secondly, from connected vehicles to automated vehicles. And eventually, a shift from automated vehicles to a coordinated automated road transport (C-ART).

C-ART is meant as an extension of the automated driving concept by adding communication capabilities that connect vehicles in between and with the infrastructure and adding a central coordination player to achieve the full potential of automated driving in terms of social, economic and environmental benefits.

The role of governments – central and local – in the evolution of automated vehicles/MaaS will be very significant in determining the progress, success (or lack thereof) of the automated driving/Maas paradigms. In this initial phase, public authorities mainly act firstly as enablers, providing the framework in which automated driving can evolve. They can keep that role, or take it further.

They may, in order to ensure that the potential benefits of vehicle automation are actually delivered, see opportunities for active traffic management by operating such services such as geofencing/control zones/emissions management (being good examples where central and local government may see active management as a part of traffic control).

As in other capacity-constrained system (e.g. aviation, railway networks), operational research highlights the need for central coordination and regulation regarding the access to optimisation and safety of any transport system. Air traffic control controls movement of aviation from the largest airliner to the smallest private plane. Readers of 'Henry the little green engine' learn in their early years that train movements are controlled by the 'big fat controller'. (If you take the AV designers paradigm where they believe the AV is in full control of its movements, transpose that paradigm to trains choosing when and where to move and at what speed, or aircraft landing and taking off from a major airport, then flying any route and height they choose, and then imagine the consequences.)

At one level, through traffic signals and signs, traffic management centres already manage and control the flow of vehicles through the major arteries of the road network, via the driver following the instructions (s)he sees. At the moment this control is effected to groups of vehicles through the means of traffic lights, VMS and physical signs. But with connectivity to vehicles, it is not such a leap to put those instructions directly into the vehicle, thus enabling the network manager to control and optimise movement of vehicles through its road network (much more about these paradigms, and C-ART, later).

One important question when assessing the potential impact on safety is how automation can address identified accident risks such as speeding or drink driving.

Speeding is cited as a primary factor in about one-third of collisions ending in death and an aggravating factor in all collisions where it occurs (although we have to be careful not to assume that speeding was necessarily the cause of the collision).

Automated vehicles will comply to static and dynamic speed limits and both car following and lane-keeping will be controlled more consistently than we would expect from a human driver and so any causational effect of speeding will be avoided. However, we cannot therefore make the assumption that the accident will have been avoided. That one or more of the vehicles was speeding may or may not have been the cause of the accident, so we should expect statistical reductions, but not elimination. This is also the case with drivers who have been drinking. We have statistical evidence that one or more drivers failed an alcohol test, but that cannot be automatically assumed to be the cause of the accident. So, again, we should expect statistical reductions, but not elimination.

Where we can be more sure, but still not certain, is in respect of accidents where the driver was using a mobile phone or arguing with another vehicle occupant. But again, we can only predict a reduction, not elimination, of accidents with these factors involved.

5

The MaaS Paradigm

We first have to be clear about what MaaS – 'Mobility as a Service' – actually means. If you lock four MaaS experts in a room together you are sure to get at least half a dozen different definitions. The difference is largely caused by difference in the objectives for MaaS as perceived by the different actors involved. And this means that the envisaged objectives, expectations and limits, are often quite different.

5.1 Purist Definition for MaaS

The best 'purist' definition that I can provide is that '*MaaS is a mindset that changes the focus of attention of the travel journey experience away from the technical means of travel (car, bus, train, bicycle, etc.) to the travel requirement of the traveller using the optimal mix of travel modes, in order to provide a cohesive, safe, efficient, comfortable, cost-efficient journey, at a specific point in time. MaaS starts where the traveller is now, and ends when the traveller has arrived at his/her chosen destination*'.

5.2 Vehicle Manufacturer Perspective for MaaS

To the manufacturer of automobiles, some may now see or be advised (McKinsey (2016)) that MaaS can redesign parts of his business model away from selling vehicles to drivers, to selling 'ride-hailing' services to provide links in a MaaS journey, or indeed the whole journey (and manufacturing/providing and operating the vehicles to provide these services [but thereby obtaining its greatest revenue income from the service provision]). Figure 5.1 (left) shows a foresight of what this may look like in the future, Figure 5.1 (right) shows an early instantiation.

This is of course a very narrow view of a market opportunity to be part of a new integrated paradigm, as a way of developing/ saving the manufacturer's business model. In this paradigm it is seen that there will be many less vehicles on the road (raising hoorahs from city planners – but not necessarily their accountants), but those vehicles on the

Automated Vehicles and MaaS: Removing the Barriers, First Edition. Bob Williams.
© 2021 John Wiley & Sons Ltd. Published 2021 by John Wiley & Sons Ltd.

Figure 5.1 MaaS concept vehicles. Source: yummybuum/123RF (left), Aurrigo Driverless Pod (right).

road will drive many more miles, so need to be replaced far more often (Ford has suggested maybe every four years).

5.3 Traditional Transport Service Provider Perspective for MaaS

From the perspective of the traditional transport service provider, public or private sector, MaaS is often viewed as using a refocus to the travellers' objectives, to ease access to and transfer between, modes of transport, increasing the attractiveness of the service provider's offering by providing a complete multimodal travel offer connecting from the first to the last mile, via a Maas 'app' that plots the routes and obtains the tickets. This paradigm envisages a MaaS service being provided usually by the local public transport provider, or by an independent MaaS 'broker' service.

5.4 MaaS from the Perspective of the MaaS Broker

Whereas the MaaS service provision is seen by public transport providers as part of the local public transport service offering, others see the MaaS service as a commercial brokerage service business opportunity, best provided by an independent organisation, because in this model, service providers do not have to share potentially confidential information with competitors.

However, in either the PT (public transport/transit) service provided model, or the independent 'broker' model, MaaS is the term used to describe digital platforms (often smartphone apps) through which people can access a range of public, shared and private transport, using a system that integrates the planning, booking and paying for travel, using a single app to acquire and pay for tickets, and is generally limited to these aspects. Figure 5.2 shows the current situation, and Figure 5.3 shows the objective.

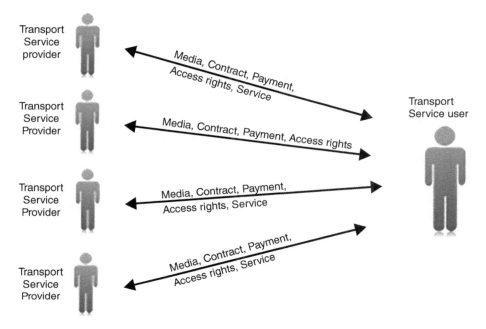

Figure 5.2 MaaS current situation.

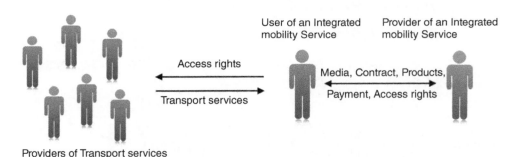

Figure 5.3 MaaS paradigm: simplification of the interfaces for the user of an integrated mobility service.

Figure 5.4 shows a view of the main services enabling the integrated mobility service. From an architectural perspective, a typical MaaS paradigm from the perspective of the PT service provider model, or the independent 'broker' model, the system architectures are similar. Figure 5.5 shows an example architecture for MaaS, while Figure 5.6 show TC204 WG19's view of the main functional responsibilities in MaaS.

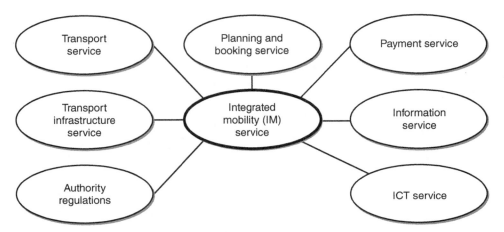

Figure 5.4 The main services enabling the integrated mobility service.

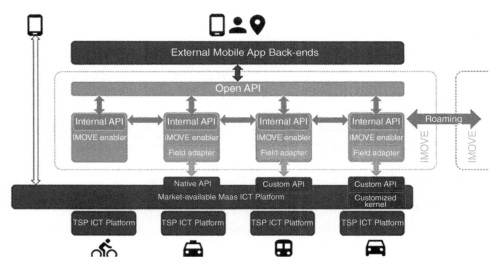

Figure 5.5 Example MaaS architecture (EC Project IMOVE). Source: IMOVE, Unlocking Large-Scale Access to Combined Mobility through a European MaaS Network, 2016, European Union.

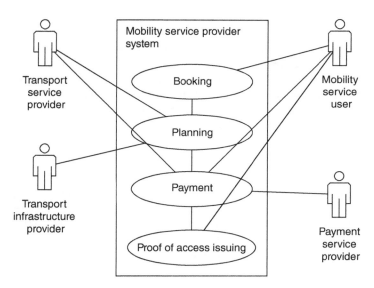

Figure 5.6 Main functional responsibilities in MaaS. Source: ISO, ISO TC204 WG19, Intelligent transport systems. © International Organization for Standardization (ISO).

The roles shown in the above diagram are explained by ISO TC204 WG19 as:

5.4.1 Transport Service

The transport service is the core service in integrated mobility as it covers the transport of a person or goods from A to B. The transport mode could be road, rail, sea or air and the transport means could be any transport facility that is used to transport people or goods. Typical examples are car, truck, bicycle, tram, train, metro, micromobility, passenger ferry and plane. Transport means also includes facilities that are not directly linked to any of the four transport modes, e.g. a gondola or a cable car in hill/mountain climbs/river crossings, etc. The transport means are usually driven by fossil or electric power, but they could also be powered by the transport user himself, e.g. bicycling or walking.

5.4.2 Planning and Booking Service

The planning and booking service is also a core service in integrated mobility. Based on the requirements and constraints of the user of the mobility service, a set of alternative integrated mobility services (often called 'products' in public transport) are presented to the user. Based on the user preferences, e.g. price, travel time and route, comfort, sustainability, transport mode and means, the user chooses his preferred alternative, books the integrated mobility service and receives the access rights to the integrated mobility service.

The set of alternative integrated mobility services are bundles of transport services linked together providing a seamless and effective travel for the user of the integrated mobility service.

5.4.3 Payment Service

The payment service enables the user of the integrated mobility service to pay for the service when the service is booked, at the moment the service is used, or after the service has been used. Hence, the service is either pre, or post,-paid. The payment could be stored in a central account or it could be stored on the payment media – which could be the same media as carrying the access rights, e.g. a smartphone or a wireless smartcard.

5.4.4 Transport Infrastructure Service

The transport means move around in a transport infrastructure built and operated for the transport means, e.g. a road network, a rail network and a traffic corridor along a coast. The transport means are depending on a safe, secure, available and effective transport infrastructure to provide the integrated mobility service with the requested and necessary quality. Hence, the building, maintenance and operation of a transport infrastructure is also a core service. Traffic management in a city is a part of this service.

5.4.5 Information Service

The information service is another core service supporting the integrated mobility service. The service includes both collection of information, information management and information distribution to other core services that are depending on information for their own service processes. Typical examples are transport service information (specification of the travel product) provided by the transport service operators and dynamic transport infrastructure information used for planning of the integrated mobility service.

The information service covers the information models and messages between the objects in the integrated mobility service domain, i.e. the 'soft' part of the information infrastructure.

5.4.6 Information and Communication (ICT) Service

The information and communication (ICT) service covers the 'hard' part of the ICT infrastructure. That means the provision and operation of the equipment required for the secure storage and transfer of information between the different services described above enabling the provision of the integrated mobility service.

5.4.7 Authority Regulations

The last domain supporting the integrated mobility is not a service, but (access to) the authority regulations defining the regulatory framework for the integrated mobility service. These could be national regulations, e.g. laws and regulations issued by the Ministry for Transport, or it could be city regulations that, e.g. define the operational prerequisites and conditions for the implementation of integrated mobility services in the city, or access control limits for various vehicle types at various times, parking restrictions, one-way traffic, etc.

5.4.8   High-Level Value Network

MaaS can therefore be described as a high-level value network for integrated mobility systems where the value network can be defined as a web of relationships that generate economic value and other benefits through complex dynamic exchanges between two or more individuals, groups or organisations. The figure below shows an example on a value network for the ITS service real-time road and traffic information. Figure 5.7 shows WG19's view of integrated mobility value networks.

5.5   MaaS as a Tool for Social Engineering

Move now to the perspective of City Authorities and Planners, Governments, and The European Commission, even the United Nations. (UN ECE Workshop on MaaS: 'As more of the world's cities become congested and polluted, new business models and technologies are emerging to solve the mobility challenge. In 2014, global venture-capital investments into mobility services amounted to more than $5 billion, up from less than $10 million in 2009.')

From this viewpoint MaaS is potentially a way to remove traffic from city centres, reduce traffic congestion, reduce pollution, achieve CO^2 reduction targets, encourage healthier lifestyles, and improve quality of life for citizens, and even countering global warming. This is a heavy social engineering paradigm, whose objectives are much more radical than simply improving the travel experience of travellers.

(UN ECE 2015) '....to contribute to the global debate and facilitate consensus on the important role of the sector in sustainable urban mobility and transport development in the 2030 Sustainable Development Agenda........... At the same time urban passenger and freight transport have negative impacts through leading to congestion, pollution and to traffic safety challenges, to mention only a few most visible pressures. Increasingly, passenger and freight movements are intertwined in a zero-sum game, ... Concrete recommendations towards the development of sustainable public transport and urban mobility networks are provided considering that all public transport modes are efficiently interconnected and that cycling and walking are integral parts of such networks as well'.

In European Commission (2017) the EC opines:

Mobility-as-a-Service (MaaS) will increasingly catalyse the public-private co-development and co-delivery of mobility and transport systems and services, as well as shared and open use of public space, data and infrastructure.

The principal prospects for decarbonisation are strong better utilisation of underused assets in transport fleets and infrastructures can accommodate increasing demand and reduce the share of unsustainable travel modes. Smart mobility systems and services have the promise to contribute to the needed decarbonisation of the transport sector and might also help address persistent problems of congestion and accessibility. However, new innovations in technologies and use need to optimise the whole transport system not road-based car travel only to make a long-term contribution to decarbonisation.

....

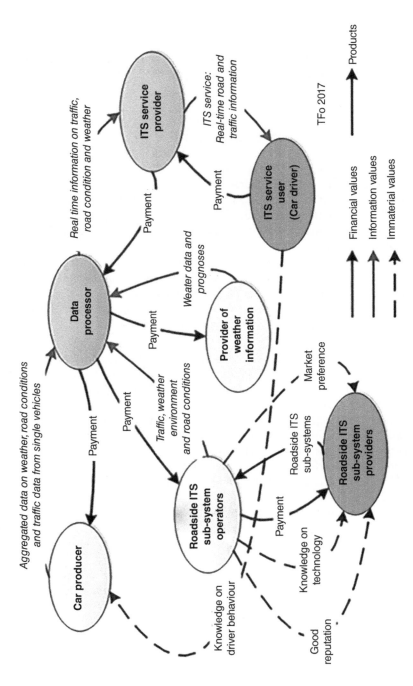

Figure 5.7 Integrated mobility value networks. Source: ISO, ISO TC204 WG19, Intelligent transport systems. © International Organization for Standardization (ISO).

Technological, socio-demographic and behavioural change are facilitating a move towards multimodal transport – combining walking, cars, buses, bikes, trains and other forms of shared transportation. Driven by the transition from 'owning' to 'using', MaaS enables multimodal mobility by providing user-centric information and travel services such as navigation, location, booking, payment and access that allow the use to consume mobility as a seamless service across all existing modes of transport.

Public and private business models, payment methods, technologies, and user choices will continue to coevolve alongside data sharing by users and public infrastructures, and increasing cooperation between the public and private sectors. MaaS should also provide more cost-efficient mobility options to consumers and households by reducing vehicle acquisition and maintenance expenditures.

.... The transformation and convergence of transport and mobility systems and services presents a unique opportunity to develop post-fossil, user-centric, smart mobility systems based on access to individual, public, shared and active mobility, rather than ownership of private automobiles.

............Integrated Mobility-on-Demand services can contribute to modal shift to public transport and also address the spatial inefficiencies of private individual motorised transport. User-centric urban mobility systems will provide ubiquitous check-in/check-out user access to enable both inter- and multi-modal mobility on demand and enhance overall transport efficiency. In future integrated and sustainable mobility-on-demand systems, electric mobility will become a component of both power and public transport infrastructure and systems. The smart integration of tariff structures, data and user interfaces as well as the disposition of rolling stock across these sectors is a central challenge, which requires new business models and scheduling, booking, navigating, ticketing and charging solutions.

..........Autonomous electric vehicles are expected to form a significant component of 'MaaS' for urban transport. As with sharing models, autonomous vehicle (in public or private service) technology will blend with MaaS models and can potentially also enable ubiquitous smart traffic management. In deploying electric and shared autonomous vehicles (SAEVs) the benefits of these mobility strategies can be combined to greater effect.

This is heavy social engineering and a world away from the more modest ambition of most MaaS service brokers. But pressures to reduce CO^2 emissions and reduce pollution will mean that there is continual and growing pressure to achieve these social ambitions. But whether or not that is in line with the purists' objective to 'improve the travellers journey experience' is another matter.

Further, the provision of even very competent MaaS 'apps' and single payment journey planning/management, by itself, does little to prize drivers out of their vehicles. But too many planners/social engineers are relying on these MaaS Apps to change behaviour, just because the MaaS apps are created.

5.6 MaaS Experience to Date

MaaS is still a developing concept. As the idea develops, so will the regulatory environment (and associated business rules) that transport must operate within. The social engineering paradigm is for the moment a dream that has yet to be realised in the context of MaaS, but it is serious European and Member States policy and policy ambition and will be driven forward.

In terms of early steps, cities around the world, and especially in Europe, are taking active steps to reclaim cities from cars, not only by charging heavily for access and parking, but by restructuring streets and public spaces to give priority to cycling and walking and as public amenity space.

MaaS is very much seen as the next stage in this paradigm. The European Commission has sponsored a number of MaaS experimental and trial projects, perhaps most notably: 'MaaS4EU', www.maas4eu.eu/project; 'Optimodal,' www.eutravelproject.eu; 'MaaSIVE', https://projects.shift2rail.org/s2r_ip4_n.aspx?p=MaaSive; 'IMove', www.imove-project.eu; 'ETC', www.europeantravellersclub.eu; 'My Corridor', www.mycorridor.eu; 'Shift2MaaS', http://shift2maas.eu, and others.

5.7   MaaS and Covid-19

Until January 2020, as we have seen above, ecologists, the European Commission, several national governments, the United Nations, many local authorities around the globe, and other agitating groups, have been flying the banner that, for the sake of the future of the planet, MaaS is the future transport paradigm: Use public transport as much as possible, with MaaS support, use multiple modes on a journey, ride share, use pickup and drop micromobility, look to the future with shared use automated vehicles, etc.

Then in January 2020 Covid-19 arrived! And in the months that followed as the pandemic swept round the world, most countries went into some form or other of lockdown.

Personal hygiene, social distancing and sanitisation became the buzzwords. We were daily lectured on the real and now existential threat of the virus. In most countries, deaths were counted in their tens of thousands, and the ever-rising number reported daily to the population. We were educated daily on how viral infection like this spreads, and as we learned more about the behaviour of the virus, we were taught to sanitise and preferably wash our hands thoroughly for at least 20 seconds after we touched anything from outside our home. Practice social distancing, wear face coverings in public places, high quality medical face masks where there known risk areas, full PPE (personal protective equipment) whenever confronted with someone who had the virus.

We watched, appalled, as the health services in many advanced countries struggled, and in some cases failed, to cope. With lockdown in place, so nothing else to do (at least in our leisure time) but watch TV that daily faced us with intense personal tragedies of those involved, and especially those who had lost loved ones and the inhumane tragedies of the situations in which they died, and the ever-mounting number of deaths.

Unlike the genteel persuasion of politicians, administrators, ecologists and 'greenies', that it would be a jolly good thing, and you know was going to be really necessary in the long run, to move to MaaS as the normal way to travel, these harsh and vivid visual images and existential threats were seared deep into everyone's brains.

And now, the very government and local authorities that had been massaging us gently towards MaaS, came out with strong messages.

Stay SAFE!

Use public transport only in an emergency.

Stay at home if you can, but if you have to travel, go by car.

Allow only your family group in your car.

Do not share cars between family groups.

Maintain social distancing at all times

Wash your hands thoroughly as soon as your journey is finished

Don't touch objects that have been handled by others, and if you do, immediately afterwards sanitise your hands with hand sanitiser. Then wash your hands as soon as possible.

Try not to touch your face with your hands when travelling.

Sanitise any objects that come into your home.

Wear face coverings in public.

Avoid cities, crowded situations, shops, etc.

Maintain social distancing at all times

Why?

Because you or your loved ones may well die, and die unpleasantly, if you do not!

These are powerful messages following what can be defined as the world's most traumatic experience since World War II.

It is true that there have been localised disasters – civil wars in many countries, local wars between countries, earthquakes, tsunamis. But globally, outside the region they occur, while we generally feel sadness and pity for those affected, we have not been terrorised by the events in the way that the presentation of Covid-19 has done, and has been designed to be, felt by all. This disease is terrible, and at the time of writing, and the time of the pandemic, has no means of prevention or known cure (written before the vaccine arrived). So, one of the few tools to fight it has been propaganda to scare the population into obeying the lockdown and social distancing. And, generally, the media around the world have done a good job of etching fear – indeed terror – deep into people's memories.

We know from the research of post-traumatic stress disorder, the effects that abnormal, violent, tragic experiences in the civil wars in many countries, local wars between countries, earthquakes, tsunamis have on those directly affected in the long term, and how in many cases this affects their subsequent behaviour for many years – even a lifetime. And we know that existential threat, especially threat to family members, scars the brain deeply, overriding logic in many cases.

So we may expect that the messages taken from the Covid-19 pandemic will be with us certainly in the short term, probably in the medium term, and in some cases in the long term. And these effects will be counterproductive for MaaS.

For many years, the use of public transport will be reduced. Businesses have found that home working can be efficient in many cases, and enable the reduction of expensive usually city based office space. In terms of saving the planet, this of course has benefit. However, not having to commute every day will start a trend away from city living. Already estate agents have seen an upsurge in interest in city dwellers seeking to relocate away from dense crowds. It has been clearly noted that small towns and villages have suffered far less from Covid-19 than cities.

The combined messages 'Stay at home if you can, but if you have to travel, go by car; use public transport only in an emergency' will push people into replacing their cars, (even more essential if you move out of a city centre) and, as we have seen elsewhere in this book, that is a long-term decision, because once you have purchased a car, the cost to use it is only the marginal cost and makes using public transport even less attractive.

And even regardless of this, the future activity of the automated car, hired by the journey, becomes much less attractive. Who was last in the car? What diseases did they have? Are the surfaces clean? This deep-seated way of thinking will take some time to fade.

As for micromobility for the last stage of the journey. Who handled the scooter/bike, etc., last? How do you sanitise it?

As these modes of transport appeal more to the young, who have been less affected, and in any event are young and less worried about existential threat, these fears will probably fade fastest, but Lime and other sharing services have already laid people off, and the recent experience certainly dampens the growth of the use of these services in the short term.

In the short term, so long as Covid-19 remains controlled only by social distancing and sanitisation, the concept of MaaS is on hold. Indeed, is unviable.

In the long term, if significant numbers of workers, freed from having to commute every day, but probably still commuting some of their time, locate themselves further into the country, this will probably increase the demand for automated vehicles, when they eventually arrive. But for MaaS, which needs growth, take up and use, in order to achieve economic viability, Covid-19 may turn out to be as big an existential threat as it is for humans.

The big question now posed for Maas is, assuming a vaccine or effective treatment within the short term, how long will it take for those deep etched fears to subside, and MaaS be once again considered as a viable way forward?

6

Challenges Facing Automated Driving

Chapter 4 described the paradigms that may evolve for automated driving, and in so doing described many of the issues that have to be faced that determine how those paradigms emerge. In summary, the challenges identified in Chapter 4 are:

(1) The first and most significant challenge facing the introduction of automated driving is that vehicle manufacturers seem to see the paradigm for automated vehicles as one where the vehicle is in sole control of its movements. But the reality is that the vehicle is just one actor in a controlled driving paradigm, and a controlled actor at that.

(2) Transport information centres, not AVs, will continue to manage the existing interaction loop between traffic information and traffic condition. Now that governments and city authorities realise the opportunities and potential benefits that Mobility as a Service (MaaS) can bring, if they are the controllers of the road network within which automated driving and MaaS will occur, *they* have now to step up to the plate and deliver the information that is required for automated driving and MaaS.

(3) Currently, automated vehicles try to know the regulations by 'reading' road signs. This is a nineteenth-century solution to a twenty-first-century problem. No matter how good camera technology gets it will never get good enough for 'safety by design'. Road regulations need to be provided/made accessible digitally to the automated vehicle.

(4) Vehicle manufacturers ploughing lone furrows to introduce automated vehicles and MaaS, using different paradigms, and defective architectures, will delay or prevent the introduction of these technologies. The 'go-it-alone', competitive strategies of vehicle manufacturers will continue to be keep vehicle technologies differentiated, and each vehicle manufacturer will implement (and keep strictly confidential) its own driving logic. In so doing are introducing considerable uncertainty and risk.

(5) New modes of transport and new mobility services must be safe and secure by design means probably by reacting tentatively after yielding, leaving generous margins to allow for the weaknesses of human drivers, or stopping, thus reducing the available space for other vehicles.

(6) What effect that weaving traffic will have on AVs? Weaving drivers tend to take significant risk, relying on the following (intruded) vehicle in the new lane to take whatever action is needed to not collide with the intruding vehicle. We know from

accident reports that in these common situations, both intruder and intruded parties take more risk than they usually consider safe, As the proportion of AVs in the traffic stream rise, this trend is expected to exacerbate the situation (congestion) even further, and will not probably improve the situation until AVs become the majority.

(7) Automation at present unfortunately does not work all of the time, and AV/autopilot mode vehicles still crash into stationary vehicles, have trouble in high contrast light situations, and sometimes don't recognise humans or animals etc. in their path, that mean that they can only be allowed to operate 'automatically' (Tesla calls it 'autopilot') under the watchful supervision of the driver, which kind of defeats the objective.

(8) Just as important as how well AVs interact with their environment, is how other road users interact with AVs. The incapability of human drivers to anticipate reactions such as the sudden stop of the leading AV due to leaves or shopping bags fluttering on the road, causes serious problems, indeed is already a cause of collisions. Time will be needed for human drivers to understand how different AVs behave and therefore it is hard to believe that AVs will contribute to safety from the first moment they appear on the road.

(9) Interaction between drivers of conventional vehicles, pedestrians and VRUs often involves eye contact, hand gestures and speech exchanges with the other party. How AVs detect such attempts at communication, and how they respond is an issue that will not only have to be thought through, but will have to react consistently if the other road user can ever trust them. Vehicles and their sensors and cameras will have to do more than simple detection and will have to be able to identify and consistently respond to different forms of communication.

(10) Project C-ART (European Commission, Joint Research Centre 2017) predicts that congestion levels will not drop significantly until AVs make up between 50 and 75% of the vehicle fleet.

(11) AVs by themselves will not necessarily be smarter than conventional vehicles driven by humans. We can assume that AVs will follow more rational rules than humans. Therefore, as with adaptive cruise control, efficiency opportunities may not be taken up by system designers, in order to better manage their risks/liabilities. They will play it safe in their designs. The adverse consequences are too costly not to do so.

(12) There has been much publicity that MaaS using automated vehicles will reduce the demand for in-city parking. And combined with policy shifts to encourage the use of public transport in cities, that seems a likely outcome.

(13) However, it will come at the expense of *increasing*, not decreasing, traffic in towns.

(14) Automated vehicles will release unsatiated demand from those unable to drive, from ageing populations and changes to commuting travel and a significant rise in additional long-distance trips could be expected, for trip distances below 500 miles (800 km) overall this could lead to a large rise in individual vehicle usage, increases of maybe 30–75%, perhaps 90%, even 160% have been predicted.

(15) Mass transit is seen by most governments to be fundamental to an efficient transport system. But a rise in the use of automated vehicles in a MaaS paradigm may increase the use of 'private' automated car journeys causing a reduction of use of public transport, threatening its viability.

(16) It is necessary that safety and some operational data must be freely shared in order to gain safety, societal and traffic flow benefits. This runs counter to automotive manufacturers strategies to monetise data from vehicles, and may also run into privacy issues.

(17) For the foreseeable future, automated vehicles will share the roads with non-equipped entities, and will have to be operable within safe and feasible contexts.

(18) With little cooperation between AV manufacturers, how AVs react to the close proximity to each other and how they react to 'incidents' and surrounding conditions is crucial.

(19) With the lack of cooperation and interworking, little has so far been said about consistency. Consistent behaviour regardless of manufacturer is crucial. And most importantly, working from the same map is also important. There are several map providers out there, and there are subtle differences in mapping methodologies between the leading providers. This results in positional differentiation. While this may not be too serious for Day 1/1.5 services, as one moves towards vehicle control in a connected paradigm, when one vehicle presents its location and direction to another, they need to be working using the same map. How this can be achieved without conferring monopoly to one provider's system is unknown.

(20) How will vehicles with speed management systems interact with unequipped vehicles? Will the unequipped vehicles travel faster and continually overtake so that the speed management system gets finally switched off by a dissatisfied driver/user? Or if he cannot do that, or it is an automated vehicle, will its journey continually be pushed back, and will the accident risk increase sufficiently because of imprudent overtakes by frustrated unequipped drivers, that the automated vehicle will slow down further and further because of the additional risk. How is that going to work out??

(21) Acceptability: European drivers can't be persuaded to hand over the simple changing of gears to automation, and most cars sold in Europe are still manual gear shift. Opinion is widespread that that AVs would not drive as well as human drivers and that users would feel uncomfortable travelling in an AV, and this has been backed up by other research. Users' acceptance towards AVs will of course grow as they become more familiar on the roads, but initial fears will inhibit early take up and affect early business models.

(22) In an environment where the communication exchange must often have low latency – i.e. have very rapid response, an agreement on a low latency communication channel is urgently overdue.

(23) In theory, the reaction time of an automated vehicle can be significantly smaller than that of a normal driver, leading to a significant increase in road capacity. However, liability issues on the responsibility of automated vehicles are likely to force vehicle manufacturers to design their vehicles to be fairly conservative, a serious problem when mixed vehicles (automated and conventional) will be on the roads together.

(24) It has been proposed that some roads could be allocated exclusively to AVs, or special roads (with very high standards of road marking), built for them. Other ideas being discussed are that some HOV (high-occupancy vehicle) lanes could be

reassigned for exclusive use of AVs, that motorways could have one lane allocated to automated and 'connected' vehicles, etc.

(25) In practice just changing a speed limit and putting up the new signs can take four to five years.

(26) Of course, automated and connected vehicles bring new requirements and challenges to the regulators in respect of traffic law, liability, security, access to data, protection of personal data, particularly for Level 4 and 5 vehicles.

(27) Political decisions need to be made as to whether the provision of data and other C-ITS service provision to vehicles from the infrastructure should be a local authority service, a regional service or a national service, or a commercial service, or a combination of all four.

(28) The role of governments – central and local – in the evolution of automated vehicles/ MaaS will be very significant in determining the progress, success (or lack thereof) of the automated driving/Maas paradigms.

(29) The operator of the infrastructure C-ITS-station will then have to identify what ITS services it will support and how. No detailed system analysis of any of these services has been circulated, and no standard service definition issued. These steps need to happen very quickly. And these services need to be defined clearly, including data tree definition and unambiguous data definition, probably in ASN.1.

(30) Fallback to driver. Where the automated system cannot reconcile its situation, in J3016 paradigms 1–4, control is passed back to the driver. Further research is needed the time needed for the driver to takeback control. What happens if the occupants of the vehicle are asleep, have been drinking, or are not qualified drivers? These are unsolved problems to this fall-back paradigm.

(31) Young drivers who have access to automated driving may build up less driving experience. This is an area that needs more research. This also poses questions for driver training: how will training teach people to drive safely and make the most of automated driving techniques, and how will drivers be taught to safely make the switch between fully autonomous and automated driving?

(32) Liability in the event of a collision with an automated vehicle needs to be more clearly defined and regulated.

(33) Cybersecurity is another major issue. In part for privacy, but more particularly to prevent hacking into vehicle control systems, or indeed the automated driving system itself.

(34) MaaS has its place, but it will not take over entirely.

The path to a world where most vehicle movements are automated are probably more than 25 years, possibly 50 or more years, away. In the meantime, the reality will be ADAS in the 2020s, with increasing amounts of control. Level 4 systems are unlikely to become widespread till the 2030s, and Level 5 maybe another decade or more behind that.

7

Potential Problems Hindering the Instantiation of MaaS

7.1 Root Causes of Obstacles

Given the very different objectives of the proponents of Mobility as a Service (MaaS), it is not surprising that there are many obstacles identified that stand before the successful instantiation of MaaS.

The obstacles identified in this chapter have largely been identified by the EU sponsored MaaS research or development projects identified in Chapter 5. To some extent they are therefore European focussed, but to our experience and observation is that they are largely generically applicable obstacles that will have to be faced in most parts of the world.

I have associated these obstacles into eight groups:

- Level of community readiness
- Level of social engineering readiness
- Perception of risks
- Level of market readiness
- Level of software solution readiness
- Training
- Timing
- Institutional and governance

7.2 Level of Community Readiness

Communities are not yet actually mature for a complete uptake of MaaS. Indeed, while it is probable that the software and commercial provisions for providing MaaS are likely to be accepted and used by current users of public transport, they may well be accepted by car drivers as a service useful for someone else, but not relevant for them.

When it comes to community readiness for MaaS as a tool to achieve social engineering, outside of government (central and local), in most countries there is little awareness of its intent, and considerable work to be done if a backlash at the ballot box is to be avoided (more of this issue in Section 7.3).

In practical terms, a large majority of journeys undertaken are mono-modal or co-modal. The development and expansion of conurbations in Europe since the middle of the nineteenth century, and particularly since the end of World War II, have

Automated Vehicles and MaaS: Removing the Barriers, First Edition. Bob Williams.
© 2021 John Wiley & Sons Ltd. Published 2021 by John Wiley & Sons Ltd.

encouraged this. The shift to multi-modality will require some significant changes in the behaviour of people (e.g. they will need to feel confident that a greater number of interchanges doesn't add risk to their travel, or be noticeably detrimental as a journey experience). This is unlikely to happen without incentive, political promotion, coercion, or force.

At a different level, MaaS is perceived largely from a city specific perspective, and for the fulfilment of local political agendas.

It is also abundantly clear that car drivers will not be coaxed out of their cars by heavy taxation. Countries like Singapore, Norway and others heavily tax car purchase (BMW 5 series price in Germany is circa €50 000, by comparison, in Norway, more than €60 000 and in Singapore over €150 000); heavily tax fuel prices, and in Singapore's case, heavily tax driving licenses. It has had little effect on car demand for the affluent, except, perversely to drive the market to larger, less fuel-efficient cars, because they have become a status symbol. The middle classes adjust the rest of their lifestyle to afford their car. But of course this is punitive and falls heavily on the poor. Drastic measures, like Norway's incentives to electric vehicles by scrapping the punitive car tax entirely, has weaned drivers to electric vehicles, but they still drive their own vehicles.

7.3 Level of Social Engineering Readiness

Through most of the developed world, and with the exception of the central densely populated core of large cities, we currently live in a culture that for most is car-centric (the exceptions being the socially needy [old, disabled, poor, etc.]). Our countries have been socially engineered to work this way over a century and more, and it by and large works reasonably effectively for those (most) people who are affluent enough to own or have access to a vehicle, with what is largely perceived as the significant benefit of a better lifestyle.

The use case is different for city centre dwellers (particularly affluent city centre dwellers) who have the benefits of a rich social city life (that is largely not shared by poor city dwellers). Here, MaaS, enhanced by micromobility options, is already perceived as beneficial, as distances are largely short, and ride-hailing/chauffeur services and excellent frequent metro links to a stop within easy walking/micromobility distance of the required destination, is readily achievable. But this sector of the population already uses these services, and so MaaS brings a (minor) improvement to the quality of life, but makes no noticeable change in travel pattern behaviour.

Some cities have focussed on changing the paradigm from expansive growth to perpendicular high rise growth (e.g. Singapore and many cities in China), and, if this limits and reduces the geographical expansive footprint of the city, will reduce traffic congestion and its associated emissions, but largely at the cost of using much more concrete (a major CO^2 emitter), and ongoing emissions from air-conditioning, delivery vans, ride-hailing vehicles, etc., and a rise in medical consequences, such as increased asthma, etc. To which the policy action is to encourage walking and cycling, and to restructure the cities to prioritise walking, cycling and public transport.

But the vast majority of cities have extensive suburbs, because we designed them that way, or allowed them to sprawl that way, and, outside of the city centre, have poor public transport options.

For example, London and the 'home counties' has a population of about 15 million, of whom 9 million live within the limit of 'Greater London', and only 3.3 million of these (22%) live within 'Central' London where there are frequent metro/bus/ride hailing routes to an extent where car ownership has only marginal benefit (Source Wikipedia). Therefore, 78% are unlikely to see MaaS as a viable option.

The Greater Paris, 'Isle de France' region, has a population of 12.2 million, of which 2.2 million (18%) live in the central (Arrondisement 75) city core (Source Wikipedia). Therefore, 82% are unlikely to see MaaS as a viable option, although it should be noted that Paris has far more extensive, subsidised public transport throughout Isle de France (especially the RER and Metro), and an ambitious Project, 'Grand Paris', to rapidly expand public transport over the next 20 years.

So cars remain an essential requirement for most citizens. Many, indeed most, workplaces have also located themselves in the suburbs to avoid city centre congestion issues, and so one (or more) car(s) are essential for most families. Having invested, by necessity, a significant amount of family income in a car or other type of vehicle, there has to be significant incentive or deterrent, to persuade the family to use another transport means for all travel, except perhaps for travel to/from the city centre. This is partly because the family has only to meet the additional 'marginal' cost of using the already purchased vehicle, and partly because (and this should not be underestimated) it represents the most convenient and comfortable means to achieve the journey, and for most journeys outside of the city centre, MaaS multimodal journeys are a less attractive option.

The willingness to change travel behaviour is very questionable. Outside of city central areas, why would a car driving traveller who can easily, and largely comfortably drive his car from start point (a), seamlessly to end point (d) and back again, elect to choose to travel from (a) to (b), wait for a connection, possibly in inclement conditions, to catch a link to (c) in order to wait for another connection possibly in inclement conditions, to catch a link to (d), and then do the same thing to return later?

However, as younger people generally have to struggle harder than their forebears to both own a vehicle, and rent or buy a home, and many also have student loan repayments to meet, and many are concerned enough about changing climatic conditions, they may be more inclined to test innovative travel modalities.

That said, there is limited evidence to support this. *The Economist* (15 February 2020) reported 'Everybody wants buses. They just don't want to use them' and detailed that Transport for London reported that the average 17- to 24-year-old took 2.3 public transport per day in 2011–2012, but only 1.7 in 2018–2019. Whereas the bottom fifth-poorest without access to a car fell from 49 to 46% in the same period, and the second fifth-poorest group without access to a car fell from 36 to 28% in the same period. Amongst the elderly, free bus pass movements also declined by around 20% despite an increase in those qualified for free bus passes.

7.4 Perception of Risks

The perception of risks and uncertainties among the different actors involved in MaaS varies considerably. And these different objectives of the different proponents make achieving success difficult, because 'success' means different things to these different actors.

For some, and particularly public transport operators, recent projects have shown that there is a reluctance, indeed fear, to put their offer in direct competition with others in a shared MaaS platform that makes evident the pros and cons for each of the options. Public transport operators are particularly afraid to lose market share to more flexible or comfortable modes.

Private operators are sometimes afraid to join a MaaS initiative because they want to be the first to instantiate this service. They can also be reluctant because they are because they believe they are responding to market opportunity/demand so need to be market responsive, but are being asked to join into a system where they have no control in decisions that are taken at a city level, or by 'brokers'.

As the chain of liability is complex for a MaaS multi-modal journey, the issue of *who* is responsible when things go wrong is complex. 'Passenger right' claiming is already a major challenge (and cost) for local public transport operators. MaaS is perceived to greatly increase that risk.

7.5   Level of Market Readiness

While MaaS is seen by planners and governments as a logical use of IT to achieve their political aims, the 'market maturity' (expectations vs. practicalities) for such concepts is questioned by many actors involved in practical MaaS projects and implementations.

Several projects have observed that user engagement is difficult when you have to deal with something so innovative, particularly where the car is seen to provide a comfortable means to undertake an end-to-end journey.

Where multiple means of transportation are available, transport operators are often working in competition with each other and there are therefore conflicts of interest between providers, which makes collaboration/cooperation difficult.

The lack of service level agreements / insurance in case of service failure (including service provision failure due to the design or operation of MaaS service software) is perceived as a significant problem.

Several projects have pointed out that current developments are too uncoordinated, and the multiplicity of projects are, in many respects, competitive with each other, hindering the wider adoption and take-up of common solutions.

Most projects reported that slow decision making had adversely affected the progress of their projects.

Early adopters (companies) are perceived to have an advantage, although if the direction of the project changes to obtain a consensus solution, early adopters may face expensive redesigns. Both early adopters and latecomers are difficult to accommodate in consensus based common solutions

The creation of the MaaS service often concentrates risk on one actor, who was most probably the 'broker' providing the MaaS service offering, who tends to be the focus for service failure, but was not in control of the transport provided (more on this in Section 7.9).

While the conceptual idea behind, and objectives for, MaaS have been thought through in many ways, and as we can see, from different perspectives, there remains a lack of clarity and consensus in working out the most appropriate business model and business rules for successful implementation of MaaS.

Some projects noted that more detailed knowledge concerning revenue handling in transport provision was needed, and lack of detailed knowledge in this adversely impacted the design of solutions.

Even where there was adequate knowledge regarding revenue handling, existing revenues models made negotiations between transport providers difficult, and made single fare solutions difficult or impossible to agree.

Difficulty on the decision making and acting as a MaaS operator (broker) has been noted by many projects, who had underestimated how difficult it would be to get the parties involved to co-operate together, and there was a general lack of understanding by the different actors involved of the different objectives of the other actors.

And several projects felt that incentives for digitization and open APIs would need to be provided.

7.6 Level of Software Solution Readiness

Current initiatives are still fragmented in terms of duration of experimentations, and features on offer.

A MaaS API is a necessity for implementation and interconnection of several functions of a MaaS ecosystem; also for shifting from one issuer to another. Alongside this, the transport corridor will need the level of connectivity necessary to support these features.

Many MaaS service providers are reluctant to share APIs, or don't have APIs to share.

There is a need for the combination and integration of the various standards and mobility token validation mechanisms and rules used by mobility service providers across Europe (e.g. Mifare, Calypso, QR Code).

Considerable work needs to be done in the area of the development of component and interconnecting interfaces, that will allow the delivery of 'traveller management' services in a MaaS.

Some elements of planned MaaS implementations will require the adoption of technology such as smart glasses and mixed reality devices.

7.7 Training

Municipalities do not yet have enough personnel trained to manage or create or operate MaaS.

7.8 Timing

There are problems with the timing of tendering processes, and also the period of contract to provide MaaS services is often too short to justify investment.

Making agreements between public and private transport service providers (car-/bike-sharing, taxis) turns out to usually be a slow and protracted process, and reluctant cooperation between competing transport means again slows the negotiation until trust is built.

Procurements by public transport authorities (PTAs) and private transport operators (PTOs) for technology platforms purchase, or development, is a slow process that can start only after an influential innovative early adopter persuades the authority to progress.

7.9 Institutional and Governance

Challenges are encountered relating to existing institutional frameworks.

There are frequent legal challenges relating to open systems, data protection, privacy and security, competition law, regulatory framework and contractual matrix and liability, payments regulation, IPR, public procurement and consumer law. These issues at least need to be addressed, and, wherever possible, practical resolution found.

Gaps exist in the competition and regulatory frameworks. For example, there are many 'automated shuttle-bus' trials involving vehicles with a few seats and some standing room for 6–12 persons. There is no UNECE category that covers these vehicles, so obtaining insurance for them is difficult. Electric scooters are banned from pavements in some countries, and from roads in others, or may simply be completely banned (as in the UK).

A governance model is needed to allow the coexistence of competitive strategies among operators that overall allow the deployment of a MaaS scheme, combining many services for the citizens, provided by a mix of public and private operators/service providers, and therefore being more attractive, partly because of competition. But competition should not be determined by price alone. Transport licencing agencies need to take quality of service issues into account when granting licences, not just price. If travellers are to be lured out of their cars it will only be for a good quality service. As Singapore, Norway and other countries who have heavily taxed car price and car use have found, demand for personal vehicles is inelastic and rather insensitive to price.

And each actor in the service provision also needs to understand and limit its liability. A clear governance model will assist such management (e.g. the public transport market in UK that is deregulated [many public transport operators]).

There is a lack of transparent 'general terms and conditions of booking and ticketing', which inhibits development, uptake and use of the service.

Private operators are often afraid to join a MaaS initiative because they provide a service to satiate an identified need, but cannot influence/take decision at city level – where decisions are more likely to be taken to satiate political needs than to satiate market requirements.

MaaS cross-border trips – different operation, payment, and ticketing policies across different cities/countries – works against successful provision of MaaS. Solutions (not yet apparent) need to be found.

Common reporting standards are required (similar to the aviation sector).

Some projects cite the lack of legislation that enforces data sharing and openness as a major inhibitor.

There is (surprising) unpreparedness of some operators to new technologies (e.g. it is often impossible to buy online tickets for public transport). For many actors a post-payment ticketing system is not feasible at this moment in time.

Unavailability of operators' data and/or data is heterogeneous. That is to say are in different formats, to different precisions, collected in different ways and are therefore difficult to assimilate.

A contractual management market place may be irrelevant for local transport as their relationships are often settled on individual approach by the provider. This is particularly true in the case of publicly owned transport services, especially where they are heavily subsidised.

Cancel and refund process for local transport, in most cases, do not exist. They need to be provided.

There is a lack of compatibility of online tickets with extant inspection and validation systems.

There needs to be improved clarity for travellers with respect to what passenger's rights apply.

Some projects cite the lack of Directives that allow public transport providers to sell services to MaaS operators as a major inhibitor.

Few of this extensive list of problems are, with good intent, good will and adoption of suitable strategy, unsolvable. However, the number of issues is considerable and the wider the combination of issues, the greater the obstacles to successful outcome. And while most of these issues can be overcome, it is clear that MaaS is not yet mature and still very much a work in progress that, in many cases faces institutional and structural changes in direction, and will take much time to maturity.

8

Potential Solutions to Overcoming Barriers to Automated Driving

Chapter 6 summarised 35 challenges – barriers – to the introduction/use of automated driving. This chapter tries to find ways to overcome these barriers and solve these challenges. And we have to be clear. Unless these barriers are overcome (by solution, adaptation, regulation, or change of perspective) no right-minded government can safely allow the use of Level 5 fully automated vehicles (AVs) in a mixed traffic paradigm.

8.1 Vehicle Manufacturers Flawed Paradigm of the Automated Vehicle

The first and most significant challenge facing the introduction of automated driving is that vehicle manufacturers seem to see the paradigm for automated vehicles as one where the vehicle is in sole control of its movements. But the reality is that the vehicle is just one actor in a controlled driving paradigm, and a controlled actor at that.

There are two potential routes to solution here.

Either automotive manufacturers will come to realise that their architecture is deficient and start to work with local authorities, transport ministries and cooperative intelligent transport systems (C-ITS) experts to work out how they can import regulations and dynamic instructions, or, regulators could enforce this as a condition of allowing their automated vehicles on the highways.

Clearly, the former route would be preferable and faster, however, it may involve the OEM's taking a pace backwards in order to make several paces forwards, and taking a pace back is always difficult. Accepting that your vehicle is just one actor among many and not the sole controlling entity is a big step to make. Possibly the threat of regulation unless AVs co-operate with road operators may be enough to encourage co-operation.

OEMs should also be encouraged to see that such co-operation can provide the last 10% that they are struggling with. AVs from several manufacturers have demonstrated their capabilities for about 90% of situations, but that last 10% is proving hard. Reading road signs with cameras will never be an adequate way to establish the regulations at any specific point of the journey (see Section 8.3 below), and 'reading' traffic lights and VMS (designed for the human eye) with digital cameras is problematic. Automotive manufacturers should instead be lobbying road operators to be pushing this information to them, or making it instantaneously available, electronically.

8.2   Vehicle Manufacturers Using Different Paradigms for Competitive Advantage

> *Vehicle manufacturers ploughing lone furrows to introduce automated vehicles and MaaS, using different paradigms, and defective architectures, will delay or prevent the introduction of these technologies. The 'go-it-alone', competitive strategies of vehicle manufacturers will continue to be keep vehicle technologies differentiated, and each vehicle manufacturer will implement (and keep strictly confidential) its own driving logic. In so doing are introducing considerable uncertainty and risk.*

With little cooperation between AV manufacturers, how AVs react to the close proximity to each other and how they react to 'incidents' and surrounding conditions is crucial.

While it is understandable that competing vehicle manufacturers want competitive segregation of their products in the marketplace, competing on the basis of 'my product's automation is better than your products automation' is a facile segregation when either parties automation only works anywhere near efficiently when they cooperate with each other on the road. Auto manufacturers, and IT companies, have come a long way since the first DARPA competition, which was designed to develop a vehicle that operated autonomously without any assistance from outside the vehicle. There has been a slow realisation that these automated vehicles have to be connected. The issue now is with whom do they connect?

Ignoring all of the work by US DoT, European Union (EU), and Intelligent Transport Systems (ITS) standards developers have made over the past 20 years, automotive manufacturers are apparently heavily influenced by the report …'Monetising Car Data- new service business opportunities to create new customer benefits' by McKinsey (2016) which advocates that

> *As privately owned vehicles become increasingly connected to each other and to external infrastructures via a growing number of sensors, a massive amount of data is being generated. Gathering this data has become par for the course; leveraging insights from data in ways that can monetize it,…*

And….

> *How might industry players in the evolving automotive ecosystem turn car-generated data into valuable products and services?*

…Through a comprehensive course of research – composed of surveys, interviews, and observations – McKinsey analysed consumer perspectives on the prospect of accessing car-generated data, identifying and assessing the value and requirements of possible car data-enabled use cases.

> *The overall revenue pool from car data monetization at a global scale might add up to USD 450–750 billion by 2030. The opportunity for industry players hinges on their ability to (i) quickly build and test car data-driven products and services focused on appealing customer propositions and (ii) develop new business models built on technological innovation, advanced capabilities, and partnerships that push the current boundaries of the automotive industry.*

And:

> *The three main value creation models of car data monetization.*
>
> *Players in this evolving automotive space are creating value from car data in one (or in a combination) of three overarching categories… First, players are generating revenue through the sale of products/services to customers, tailored advertising, and the sale of data to third parties. Second, they are using car data to reduce costs by, for example, making R&D more efficient or minimising the need for repairs. Third, players are increasing safety and security by leveraging car data's ability to speed up safety interventions that protect drivers from physical harm or from the theft of their belongings or personal information.*

… Have favoured a concept called the 'extended vehicle' (ExVe) whose concept is that the vehicle passes its generated data to the proprietary 'cloud' of the vehicle manufacturer who will manipulate and process the data to sell as value added services to the car owner, and, aggregated, sell to other third parties.

This model, favoured by automotive manufacturers, is predicated on the basis that because the data is generated by a vehicle they manufactured, and is automatically, and largely without the vehicle keeper/user's knowledge or consent, passed to the manufacturer's cloud, therefore the property of the vehicle manufacturer to use, manipulate and sell.

While enthusing about the opportunities to monetise

> *we identified more than 30 distinct use cases that hold the potential to monetize car data,*
>
> *Laying the groundwork to develop and, thus, monetize these experiences will require investments in car technologies, such as cameras to capture the driver's image and other sensors, as well as in a communications platform. Multiple players can cooperate and compete in this arena,*

etc., there is very little said in the McKinsey report about data ownership, or data privacy.

To be fair, the report does say

> *Regulators/government institutions are setting the standards regarding the collection and sharing of car data. They are also in a position to mandate car data-enabled services that support the public good, such as emergency call features, and regulate controversial topics, such as technical certification of the connected vehicles, data ownership rights, and intellectual property rights over shared technologies and services…..*

…. and ….

> *Tackling the issues related to the concept of data ownership, the rights of database owners, data anonymization/pseudonymization, and mandatory data security*

> *measures will be critical in accelerating (or hindering) the development of the use cases described above.*

But automotive manufacturers, speaking on several occasions in TC22 meetings with others, with the EC, and through OICA at UNECE WP29, constantly maintain that 'as it is the heavy investment by automotive manufacturers that has made this data possible and the data is collected to constantly improve the performance of the vehicle and thus benefit the vehicle user, it is only right that the manufacturer can monetise this data to repay its heavy investments' or similar words.

This argument is specious. If you generate data on an Excel spreadsheet and store it in an Access database on your desktop or laptop; because HP, Samsung, Apple, or whoever made your laptop/desktop/phone had to invest heavily in the technology to enable Excel or Access to operate on your machine; or because Microsoft invested heavily in the Excel or Access software, does that mean that HP, Samsung, Apple or Microsoft can claim ownership of the data you generate? Of course not.

Data is not generated by the vehicle, it is generated by the person using the vehicle, and, in Europe at least, under GDPR (General Data Protection Regulation (TFL report 2015/16) is 'personal' data (because it can be associated with the data generator), and not in the ownership of, but under the protection of the data controller/data processor, who is bound to respect the ownership, privacy and use of personal data. The McKinsey report does not mention GDPR once, even though it was highly topical at the time the report was produced.

This author is not sure how (statement in the McKinsey report)

> *We will tackle the technological enablers in detail in the following, but it is worth noting how gaining access to car data and being able to link the data streams to the ID (identification code) of the customer who generated the data, are critical elements to monetize*

envisions compliance to GDPR, but it seems unlikely.

To make matters worse, each of the major automotive manufacturers is pursuing its 'go-it-alone', competitive strategy in the way that it instantiates its Ex/Ve implementation, and, while being bullish about the potential, recent tests between the association of European Insurers and the leading five vehicle manufacturers in Europe demonstrated that not one could properly and securely make its ExVe work, and one conclusion was that it would only be feasible once 5G communications were widely deployed across Europe.

Where do we go from here?

There are only two options:

- either the vehicle manufacturers have to change direction and cooperate better on exchanges of vehicle data (probably via UNECE WP29, where vehicle manufacturers and participating governments are both represented), or
- the EC and US DoT will have to regulate regarding
  a) Data ownership
  b) Data required for safety services
  c) Data Access
  d) Cybersecurity

What is certain is that connected and automated vehicles cannot successfully operate and share data using proprietary and processed data from proprietary ExVe networks. Data must be exchanged directly >< between vehicles, or >< between the infrastructure and vehicles. The data exchanged must be freely available within cybersecure transactions, and must be provided with very low latency. ExVe does not satisfy that paradigm.

To be clear, so long as the genuine freely obtained consent is obtained (and under GDPR, consent cannot be a precondition of provision of a service) there is no objection to ExVe being used to assist operation or maintenance information concerning the vehicle, (in Europe, so long as that data/information is available under the access rights conditions of the single open market), but it is not suitable for C-ITS communications.

Both options are slow to pursue. This does not matter too much for automated vehicles because full automation is, as we have seen, probably a decade or more away, but the interim J3016 Levels 1–3 urgently require this data access to facilitate provision of these services, immediately. In the EU and its aligned countries, this probably requires a Delegated Act, as a matter of urgency, and whilst US DoT does not have such extensive powers, probably an Advisory Notification of some description.

8.3   Road Operator's Responsibilities

Transport Information Centres, not AVs, will continue to manage the existing interaction loop between traffic information and traffic condition. Now that governments and city authorities realise the opportunities and potential benefits that MaaS can bring, if they are the controllers of the road network within which automated driving and MaaS will occur, they have now to step up to the plate and deliver the information that is required for automated driving and MaaS.

Currently, automated vehicles try to know the regulations by 'reading' road signs. This is a nineteenth-century solution to a twenty-first-century problem. No matter how good camera technology gets it will never get good enough for 'safety by design'. Road regulations need to be provided/made accessible to the automated vehicle.

This is the other side of the coin to Section 8.1. It is not enough to consider automotive manufacturers automated vehicle architectures and paradigms inadequate. OEMs can only use regulatory information and dynamic information if it is provided. Therefore, traffic signals need to have an ITS-station capability to tell oncoming vehicles their status and status changes.

Importantly, vehicles need electronic access to traffic regulations:
– the need:

- Means to encode and exchange traffic regulations
- To support modernisation of the digital transport economy
- Provide critical source of data for C-ITS Connected & Automated vehicles (CCAM)

– the problem (Figures 8.1–8.3):

No matter how good your camera technology, no matter how clean your camera lens, no matter how good your system software, it will not be able to read reliably in any of the situations below, and other examples and there are plenty more examples that we could show.

Figure 8.1 Roadsigns that confuse CAVs. Source: Bob Williams.

Road signs are also often hidden behind high-sided vehicles, damaged, vandalised, or contain small textural conditional information that can't be read, or interpreted on the fly, by a moving vehicle (Figure 8.4).

Best guess is that it will never be possible to 'read' more than about 90% of road signs, often less. And that is just not good enough for a safe AV paradigm.

But trying to read road signs is a nineteenth-century solution to a twenty-first-century challenge.

In the 1800s road signs started to be erected by LAs as the only method to tell (horse-drawn) carriage drivers what the regulation was at that place...... Metal signs were seen as the way forward... the new industrial age technology way .... for direction and regulation signs (replacing engraved stone)

Figure 8.2 Roadsigns that confuse CAVs. Source: Bob Williams.

But trying to get vehicles to read signs is treating the symptom not the problem.

An analogy: The milkman used to deliver milk to the doorstep using a horse and cart daily. Today you do the weekly shop online and the supermarket delivers it to your 'fridge' .... What use/relevance is replacing the milkman's old horse with a younger one?

The issue is not about reading signs, but about getting accurate regulation information to the vehicle navigation and control systems.

If you choose to do this using deficient nineteenth-century technology as part of the system, you are doomed to failure, and a read rate too low for you to ever safely allow AVs onto shared roads..........it is that stark a choice!

In this digital age, the same way as all information is transferred these days, vehicles/map providers need that information to be provided digitally.

Figure 8.3 Roadsigns that confuse CAVs. Source: Bob Williams.

Figure 8.4 Roadsigns that confuse CAVs. Source: Bob Williams.

Figure 8.5 Current road regulation written a long time ago. Source: Metropolitian police Traffic regulation ©Metropolitan Police Authority.

And yes. This is a big challenge because most regulations are written (often imprecisely) on paper and stored in a filing cabinet (Figure 8.5).

But the regulators, will never allow AVs onto shared roads unless the vehicles have accurate information............ ....not 60–80%, not 90–95%, but better than 99.9%.

This requires 'Management of Electronic Traffic Regulations' (METR).

In 'Simple' Terms, Regulations

- General rules of the road embedded in national/regional laws, generally aligned to international conventions, e.g. default speed limits; driving side of the road
- These are often unsigned, or partially signed
- Traffic signs, signals and road marking – communicating obligations, warning or information
- Impacting: moving 'traffic' (i.e. typified road users), non-moving traffic

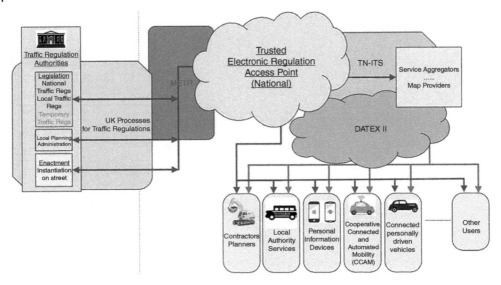

Figure 8.6 METR architecture. Source: ISO TC204 WG19 working document.

- Both 'static' and 'dynamic' (e.g. road works, road closures, dynamic speed limits) (Figures 8.6 and 8.7)

METR – Key Concepts

- A means to encode traffic regulations electronically to be machine read, processed, and correctly interpreted
- This is challenging!! Means to encode traffic regulations (noting the challenge given the variety we have even nationally, and the huge variation that will exist across nations). These specifications will be to be very specific on matters of location (there has been discussion of a single digital map for a jurisdiction) – then think of map edge matching
- A means to securely, exchange data in a traceable manner
- Security, authentication, certification – essential to trust in the solution

The question, then arises 'but hasn't this already been done?'

DATEX II

- Provides a partial fit
- But is designed for the traffic management/traveller information exchange paradigm
- It describes dynamic characteristic (events or attribution) using various location referencing methods against no-specific reference network (but dynamic features might never change!)
- It should be noted that extension work is ongoing

TN-ITS

- Is a partial fit

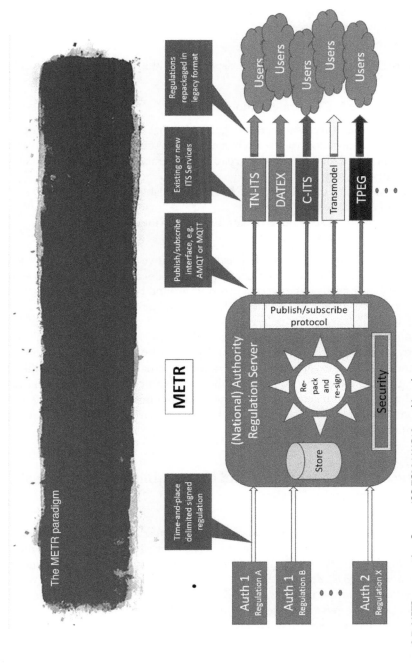

Figure 8.7 METR perspective. Source: ISO TC204 WG19 working document.

- TN-ITS is designed to pass road network attribute changes to map suppliers
- It uses various location reference methods against no-specific reference network (but perhaps INSPIRE networks offer a fall back)
- Extension work is ongoing
- UVAR – Urban Vehicle Access Regulations
- GDF INSPIRE/TC211

All of these initiatives cover different aspects, but the problem is how to fit this together?

The Outcome of the METR Project Should Be

- A set of technical specifications and guidance that support the definition and deployment of a common robust exchange of road traffic regulations between the definitive sources of that information (road administrations) and a variety of users in a machine-centred way
- Agreement and list of data concepts to be made available
- Definition of data concepts (in ASN.1)
- Definition of standardised means of data access/distribution
- Definition of standardised means of data access transactions

The recommendation is therefore that regulators and vehicle manufacturers need to work together for mutual benefit. Standards need to be developed for common transactions. Cybersecurity needs to be agreed, probably around ISO TS 21177 – an enhancement and extension of the model that the EC Joint Research Centre (JRC) has trialled and tested, and published as EU policy for C-ITS

Most importantly, national governments need to determine how they will provide this data/transactions (e.g. at national, regional or local authority level, or combination thereof) and need to press ahead with their local instantiations of METR/unified traffic information system access points. This data needs to be available *before* Level 4/5 automated vehicles can be allowed to hit the road, and preferably by the time Level 3 automation is available. So, government and local authorities need to start developing and implementing *now*.

8.4 New Modes of Transport and New Mobility Services Must Be Safe andSecure by Design

New modes of transport and new mobility services must be safe and secure by design means probably by reacting tentatively after yielding, leaving generous margins to allow for the weaknesses of human drivers, or stopping, thus reducing the available space for other vehicles.

What effect that weaving traffic will have on AVs? Weaving drivers tend to take significant risk, relying on the following (intruded) vehicle in the new lane to take whatever action is needed to not collide with the intruding vehicle. We know from accident reports that in these common situations, both intruder and intruded parties take more risk than they usually consider safe, As the proportion of AVs in the traffic stream rise, this trend is expected to exacerbate the situation (congestion) even further, and will not probably improve the situation until AVs become the majority. Automation at present unfortunately does not work all of the time, and they still crash into stationary vehicles, have trouble in high contrast light situations, and sometimes don't recognise humans or animals, etc., in their path, that mean that they can only be allowed to operate 'automatically' (Tesla calls it 'autopilot') under the watchful supervision of the driver, which kind of defeats the objective.

As described in Chapters 3, 4, and 6, there are questions about how automated vehicles handle proximity to human drivers and other impediments that they encounter, and there are questions about how automated vehicles react to actions of other automated vehicles that are operating to the system of a different operator (with maybe different logic, control parameters and behaviour). Insurance and liability issues make it almost certain that they can only be programmed to react with generous safety margins, the consequence of which will be reduced road capacity and a consequent increase in congestion. As we have seen above, there are also cautious responses that will encourage human drivers to queue-jump, pushing the automated vehicles ever further back in the queue, no doubt to the frustration of the users of the automated vehicle. Will this deter them from using an AV in the future? This can only be a matter of speculation at the moment.

How to overcome this potential barrier?

Fortunately, before the advent of Level 4 and 5 vehicles on the streets we likely have a few years of 'connected' vehicles, where vehicles, of different marques will be communicating with each other, and manufacturers can use this period to ensure that there are adequate exchanges so that behaviour between AVs of different marques can be known/predicted without overcautious response.

The reaction to human behaviour will be more difficult to overcome. In addition to the human driver taking advantage of what it knows must be the response of the automated vehicle, and therefore taking greater risks queue jumping because the driver knows the AV will make an emergency slow down, there is also, given the same knowledge, the risk of pedestrians walking out in front of an AV, say to cross the road, because they can rely on it making an emergency stop to miss the pedestrian.

It is not easy to find a solution to these problems and national governments will probably have to revise their dangerous driving and jaywalking regulations to make such behaviour explicitly illegal.

8.5 How Other Road Users Interact with AVs

Just as important as how well AVs interact with their environment, is how other road users interact with AVs. The incapability of human drivers to anticipate reactions such as the sudden stop of the leading AV due to leaves or shopping bags fluttering on the road. Time will be needed for human drivers to understand how different AVs behave and therefore it is hard to believe that AVs will contribute to safety from the first moment they appear on the road.

The solution to this issue is, as the text indicates 'Time will be needed for human drivers to understand how different AVs behave'. The penetration of AVs will be slow, allowing plenty of time for human drivers to learn to live alongside AVs, and learn their different behaviour foibles. We have all learned quite quickly that most BMW drivers drive more aggressively than most Volvo drivers and adjust our reactions accordingly. Time and the experience gained with that time should solve this problem.

8.6 Automated Vehicles Will Have to Be Able to Identify and Consistently Respond to Different Forms of Communication

Interaction between drivers of conventional vehicles, pedestrians and VRUs often involves eye contact, hand gestures and speech exchanges with the other party. How AVs detect such attempts at communication, and how they respond. Will not only have to be thought through, but will have to react consistently if the other road user can ever trust them. Vehicles and their sensors and cameras will have to do more than simple detection and will have to be able to identify and consistently respond to different forms of communication.

It is reasonable to expect an AV to tell the difference between another AV and a vehicle driven by a human driver. Hopefully this will be achieved by car to car C-ITS communication, but in any event, intelligence in the vehicle will probably be able to detect another AV by its behaviour. AVs will of course detect pedestrians, and hopefully, cyclists and for insurance and liability reasons, may be expected to act cautiously and defensively. Whether or not the pedestrian or cyclist is gesticulating or shouting is unlikely to be detected, but sudden movements will be detected. At the time of writing, problems still exist between identifying the difference between small children and animals. Problems still exist identifying items floating in the wind (such as empty plastic bags, newspapers and spirals of leaves moving in an eddying current of wind). However, these are largely seen as 'objects' to be avoided, and the AV reaction is to stop or take evasive action. The biggest downside of this is if the object appears close to the vehicle causing an emergency stop, there is serious risk of read-end collision from a following vehicle.

Given the speed of development of intelligence in vehicles, and a probable roll-out of Level 4 and 5 vehicles being more than 5 years away, probably nearer 10 years, it is not unreasonable to expect the vehicles artificial intelligence (AI) to have improved enough to settle these issues before roll-out occurs. What is clear at the moment, however, is that today's test vehicles, rolled out automation/autopilot features are inadequate at the moment to allow automated driving without a watchful driver in the vehicles supervising and ready to take over at any time. In the face of the evidence of several –many – (and often fatal) crashes that have occurred on the roads in the period 2017–2019, this author questions the authorities' decisions to allow vehicles with these features onto the roads.

Example: The 23 March 2018 crash on highway 101 in Mountain View. *Associated Press, The Guardian,* and others, showed pictures of the scene where a Tesla electric SUV crashed into a barrier on US Highway 101, California. *The Guardian* newspaper reported that 'Tesla has said a car that crashed in California last week, killing its driver, was operating on Autopilot' (Figure 8.8).

A crash in Arizona in 2018 was the first recorded AV accident that killed a pedestrian and showed some weaknesses in the law that need to be remedied. The crash involved an adapted Volvo XC90, being tested by Uber. The pedestrian was pushing a bicycle laden with shopping bags. The vehicle was operating in autonomous mode and the car's human safety backup driver could not intervene in time to prevent the collision. Vehicle telemetry obtained after the crash showed that the human operator responded by moving the steering wheel less than a second before impact, and she engaged the brakes less than a second after impact (NTSB, 2018).

The National Transportation Safety Board (NTSB) (2018) preliminary findings were substantiated by several event data recorders (EDR) and proved the vehicle was travelling 43 miles per hour (69 km/h) when the victim was first detected six seconds (378 ft) before impact; during four seconds the self-driving system did not infer that emergency braking was needed. According to *Wikipedia* (2019) a vehicle travelling 43 mph (69 km/h) can generally stop within 89 ft (27 m) once the brakes are applied.

Human intervention was still legally required. But from the moment that the take-back control alert was sounded, to the impact, provided inadequate time for the safety backup driver to respond and stop the vehicle. This raises several legal issues that need to be clarified by new regulation.

The first issue is that for Levels 1–3 (and seemingly some L4 situations) it is predicated that the driver must be able to take back control of the vehicle when required.

The UNECE WP29 document 'Revised Framework document on automated/ autonomous vehicles' (2019) states 'Automated/autonomous vehicle should include driver engagement monitoring in cases where drivers could be involved (e.g. take over requests) in the driving task to assess driver awareness and readiness to perform the

Figure 8.8 AV crash. Source: Guardian newspaper.

full driving task. The vehicle should request the driver to hand over the driving tasks in case that the driver needs to regain a proper control of the vehicle.'

The document 'Legal_consequences_of_an_increase_in_vehicle_automation, Part 1' Bundesanstalt für Straßenwesen (n.d.) states: 'However, for the safe use of the system it is still always required that the driver constantly monitors the system and understands the limits of the system as well as his own abilities well enough to realise when a correction of system control is required. System limits will persist and errors may occur – as explained above. The partially automatic system therefore frees the driver only from the active execution of action – until he regains the necessary active control in driving the vehicle.'

The Uber crash was part of a trial with a professional safety backup driver (not, for example, a vehicle on autopilot with the driver prioritising and occupied with other things). The car was not driving at high speed. Nevertheless, the safety backup driver's reaction time was inadequate to prevent the crash. This brings into question the whole assumption of '– until he regains the necessary active control in driving the vehicle.' And it shows that while it is normally a practicable action in a non-emergency situation for the vehicle to report it has a problem and alert the driver to take back control, in an emergency (likely crash) situation, however, alert the driver, (s)he will almost never have the reaction fast enough to avert the crash. This assumption that the driver overseeing the vehicle operating in automated mode can take back control in the event of an emergency therefore needs to be challenged and legally revised before a*ny* level of automation is allowed, unless the failsafe provision is proven to be adequate.

Bundesanstalt für Straßenwesen (n.d.) states: 'As for the return of vehicle control to the driver, further consideration must be given to the fact that the fully automated system is able to automatically return to the safe state; this means that no further dangers arise if the takeover does not happen.'

Secondly, neither the vehicle manufacturer, the designer of the automated system, nor the operator of the vehicle (Uber), could be prosecuted. (Uber responsibly stopped its trial after the accident). The law in every country in the world says in one way or another that 'the driver is in control of the vehicle at all times'. So, the only person that could be prosecuted is the poor safety backup driver, who was in a situation she *could not* control.

This has to be changed in law. If an automated vehicle kills or injures a person, the designer/operator of the system must be held to account by the law. They may or may not be found guilty, but their responsibility must be testable under the law. If this change does not happen, there may well be a backlash against AVs.

Perhaps early action with a recommended regulation developed by UNECE WP1 and/or UNECE WP29 could provide guidance to nation states as to how to amend their regulations.

8.7 AVs by Themselves Will Not Necessarily Be Smarter than Conventional Vehicles

AVs by themselves will not necessarily be smarter than conventional vehicles driven by humans. We can assume that AVs will follow more rational rules than humans. Therefore, as with adaptive cruise control, efficiency opportunities may not be taken up by system designers, in order to better manage their risks/liabilities. They will play it safe in their designs. The adverse consequences are too costly not to do so.

With all of today's hype about artificial intelligence and the so-called approach of 'singularity', where artificial intelligence takes over the leadership from mankind, it is easy to be persuaded that tomorrow's car of the very near future will be more intelligent than the human driver. But based on the decade of experience of the development of the automated vehicle, and the millions of miles driven to get it automated driving right and practicable, and not getting there yet, we are nowhere near this point yet.

Further, as well as perform the tasks efficiently, AVs have to perform in a manner that follows the expectations and comfort of its passengers.

Although artificial intelligence (AI) based on deep learning architectures using deep neural networks are being used in AV projects Langenwalter (2016), such systems are being used to replicate human thought and reasoning, (rather than develop alternative paradigms). This is because AVs are designed to be used in an existing (transportation via a road network) paradigm, and are not being used to replace road transport with a totally different travelling paradigm (e.g. transport by drones).

AVs use artificial intelligence and deep learning capabilities to make informed decisions and discern their surroundings – emulating the human brain functions. Deep learning can be used to emulate the way the human brain learns about the world, recognising patterns and relationships, understanding language and coping with the ambiguities encountered in life, and, particularly in this paradigm, in the process of driving on the road network.

AI techniques also enable the system to learn from their collective experience, not just within one AV, but from a network of communicating AVs. Unlike human drivers, automated driving systems (ADSs) will not get distracted or tired. They will operate more systematically than human drivers and may be able to develop some 'intuitive' solutions to situations encountered, but there is little evidence as yet that they may be able to react more quickly. They will react more consistently, but there is not yet any evidence that they will react more cleverly. Indeed, they will be in the process of 'catching-up' for some years to come.

The most common use of artificial intelligence/machine learning in AVs is for object recognition and distinguishing between objects, and improving the information received from radar, lidar, and cameras, and combined use of these technologies to identify patterns in data, without being explicitly programmed. The technology is also used to manage passage through predicted congestion.

AI machine learning techniques are being used because the spatial resolution of radar is typically comparably poor (Okumura et al. [2016]) while camera systems provide high-resolution 2D images, whereas high-definition lidars typically give lower-resolution 3D images and information. All areas where human drivers usually excel.

The consequence is that traffic planners, and road operators therefore need to understand that AVs might better systemise the use of roads, but in the near and medium future, that systemisation will reduce road capacity. Arguments and investment cases that use the 'intelligence' or the 'systemised' approach of AVs leading to improved capacity of the road network should be avoided.

8.8 Congestion Levels Will Not Drop Significantly

Project C-ART (n.d.) *predicts that congestion levels will not drop significantly until AVs make up between 50% and 75% of the vehicles fleet.*

As we have seen in the discussions in the earlier chapters of this book, while AVs (assuming they are widely adopted) in a MaaS context, will reduce the numbers of vehicles in the car park, they will be spending a far higher percentage of their time in traffic on the road. We have significant evidence with the ride-hailing services (such as, but not only, Uber), that the use of private vehicles declines in the presence of ride-hailing, but congestion is exacerbated by the additional movement of these vehicles between fares Sabur (2017). (The only principal difference between ride-hailing and automated MaaS vehicles is the presence of a driver).

Evidence from San Francisco also indicate that these services are increasing vehicle miles travelled in dense urban areas. In San Francisco, where a recent study *San Francisco County Transportation Authority* (2018), *TNCs & Congestion,* www.sfcta.org/emerging-mobility/tncs-and-congestion attributed half of the observed congestion growth from 2010 to 2016 to ride-hailing companies.

In the UK, as part of DfT's Road Traffic Forecasts, tests were conducted to investigate how travel demand might change with the deployment of connected and self-driving vehicles. These suggested that in some scenarios traffic could grow significantly Future levels of demand and congestion will be highly dependent on whether ride-sharing is widely adopted. As this author opined earlier, the issue of acceptance of ride-sharing is a social issue, and is not connected to either automated driving or MaaS, but it is probably worthy to note that UK DfT recently acknowledged that mobility innovation must help to reduce congestion through more efficient use of limited road space, for example through sharing rides, increasing occupancy or consolidating freight. *There is finite road and pavement space in our towns and cities, many of which were laid out long before the advent of motorised transport. The lower running costs enabled by new technologies and business models could worsen congestion if vehicle occupancy and load factors remain low.*

Todd Litman of the Victoria Transport Policy Institute in *Autonomous Vehicle Implementation Predictions Implications for Transport Planning* (2017) predicts that some benefits, such as independent mobility for affluent nondrivers, may begin in the 2020s or 2030s, but most impacts, including reduced traffic and parking congestion (and therefore road and parking facility supply requirements), independent mobility for low-income people (and therefore reduced need to subsidise transit), increased safety, energy conservation and pollution reductions, 'will only be significant when autonomous vehicles become common and affordable, probably in the 2040s to 2060s, and some benefits may require prohibiting human-driven vehicles on certain roadways, which could take longer.'

Project C-ART opines that 'The increased road transport demand, which is expected to arise with highly automated vehicles, may at some point exceed the available road capacity, thereby potentially leading to congestion peaks, with severe consequences. The role of transport authorities in managing transport automation in real time can become crucial. After public authorities' initial enabling efforts, a totally different role can be played, in order to ensure that the potential benefits of vehicle automation are actually delivered with automated vehicles integrated in the management of the whole transport system.' We will return to the consideration of Project C-ART later.

To further complicate the issues, and as discussed in Chapters 3,4, and 6, it is likely that the ability to work and rest while travelling may induce some motorists to choose larger vehicles that can serve as mobile offices and bedrooms and drive more annual miles. Although the additional vehicle travel provides user benefits (otherwise, users would not increase their mileage) it will possibly significantly increase congestion and may increase (tailpipe and non-tailpipe) pollution emissions.

If the advantages of platooning are seen to be great enough, some lanes, possibly some roads, may be dedicated to platoons, so non-automated vehicles may increase congestion on alternative highways and backstreets. If the use of automated vehicles in a MaaS paradigm, as discussed in Chapters 3,4, and 6, reduces public transit travel demand, leading to reduced service, the consequence will be to stimulate more sprawled development patterns, which increase total vehicle travel.

The conclusion at this stage, with the little real experience/evidence we have today, is that overall, automated vehicles/MaaS are unlikely to produce any reduction in road loading or congestion before the 2040s at the earliest, possibly the 2050s or later, and that in the 2030s assuming Level 5 becomes practicable by then, may increase traffic loading/congestion.

In the meantime, automated driving and MaaS justifications for investment and use should not be based on traffic and congestion reduction, except perhaps regarding investment in shuttles on segregated or limited speed roadways, where AVs can be used to park conventional vehicles away from town centres, or avoid the use of private vehicles altogether.

8.9 Automated Vehicles Will Release Unsatiated Demand

Automated vehicles will release unsatiated demand from those unable to drive, from ageing populations and changes to commuting travel and a significant rise in additional long-distance trips could be expected, for trip distances below 500 mi overall this could lead to a large increase of maybe 30–75%, perhaps 90%, even 160% rise in individual vehicle usage have been predicted.

As explained in Chapter 4, many studies have estimated a potential increase in the number of vehicle kilometres travelled as a result of these new technologies. 'The increased demand is both because existing latent demand from these new (underserved) groups of users can now be satiated, and the creation of new demand resulting from capacity improvements enabled by AVs More attractive travel conditions, and the fact you can work during the journey, will encourage longer commutes, thus further increasing road miles driven.'

These forecast figures are very significant and this is primarily because regarding currently underserved users:

- The population is ageing, with currently (in the United States and western Europe) a little over 15% being over 70 years old. This is predicted to rise to up to 30% in a decade's time. Already, the number of Japanese people over 65 years or older has nearly quadrupled in the last 40 years, to 33 million in 2014, accounting for 26% of Japan's population. Automated driving is a way for this segment of the population to retain its independence

- At the other end of the scale, the proportion of young people who can drive is decreasing. This is mainly due to student debt, the increased difficulty in affording to get on the housing ladder, and a significant increase in the number who choose to, and are encouraged by governments to cycle, especially in towns and cities.
- Driving is often not a possibility for the disabled, in part possibly because of the nature of their disablement, but mainly because of the cost.
- After many years of austerity following the 2008 banking crashes, there is a growing proportion of people who cannot afford a car
- There is a generational difference with young people being far more used to obtaining services through 'apps' on mobile phones.

And the attraction of a longer commute cannot be understated: We have direct evidence of this from the expansion of the London underground in the 1920s and 1930s. The period between the two World Wars saw London's geographical extent growing more quickly than ever before or since. This was facilitated by expansion of the underground railway system (running largely on the surface to new suburbs), with a significant expansion of the rail network (already extensive) to the south. Widening car ownership further advanced this trend of a preference for lower density suburban housing, by Londoners seeking a more 'rural' lifestyle, with a manageable commute. Similar trends occurred in the United States with the expansion of the federal highways network and an expansion of urban highways.

Today, with heavy traffic, and overcrowded trains, the commute is seen as an unpleasant necessity, wasting probably two hours of unpleasant and unproductive time, sometimes significantly more. If automated driving changes that paradigm to one where you travel in a comfortable AVs with the possibility to work, or lounge, relax, watch films, it is difficult not to envision the popularity of such a commute. In a world of MaaS automated vehicles this could significantly increase the load on the road system. For some and especially with 5G communications, the commute could simply become part of the working day.

Finally, many drivers are not keen on driving more than about 250 mi per day. Fatigue, tiredness and stress being major factors, all of which are removed in the AV paradigm. As stated in Chapter 4, AVs offer a viable alternative for land journeys, thus further increasing load on the road network. La Mondia et al. (2016) in a paper to the Transport Research Board in the United States, postulated that, a significant rise in additional long-distance trips could be expected with AVs, for trip distances below 500 mi.

The Fraunhofer Institute Dungs et al. (2016) commissioned a survey of 1500 motorists as part of the 'Value of Time' study. Their findings included that motorists were prepared to pay more for AV services. The more popular automated driving becomes, the greater the demand by users for services to meaningfully utilise the time freed up in the car.

Further, in comparison to self-driven commutes, Le Vine, Zolfaghari and Polak (2015) argue that because vehicle passengers tend to be more sensitive to acceleration than drivers, and occupants use travel time to work or rest it is plausible that for comfort sake users will program their vehicle for lower acceleration/deceleration characteristics than human-powered vehicles, further exacerbating the adverse effect on total urban roadway capacity.

If connected and automated vehicles lead to a greater travel demand and preference for car ownership, Project C-ART (of which more later) posits that a totally different management of the road transport system might be required. From higher levels of coordination up to the complete control of the system might be required to ensure that the performances of the road system will not gradually deteriorate or even collapse.

Project C-ART concludes that as soon as the share of vehicles with higher degrees of automation become substantive, the need for different approaches to traffic monitoring and control will immediately emerge.

Road operators, planners, and governments need to realise at an early stage that the cumulative effects of AVs may be beneficial, in some cases extremely beneficial, but it is likely to come at the cost of increased demand on the road network, and is extremely unlikely to reduce road miles driven. Statements like 'Advocates predict that autonomous vehicles will provide significant user convenience, safety, congestion reductions, fuel savings, and pollution reduction benefits' should not be trusted, and should be challenged. The benefits of AVs lie in serving the community better, not in reducing congestion, or fuel savings, nor pollution reduction.

8.10   Safety and Some Operational Data Must Be Freely Shared

It is necessary that safety and some operational data must be freely shared in order to gain safety, societal and traffic flow benefits. This runs counter to automotive manufacturers strategies to monetise data from vehicles

Automotive manufacturers, carried away with the McKinsey report, 'Monetising Car Data- new service business opportunities to create new customer benefits' (2016) have favoured a concept called the 'extended vehicle' (ExVe) – see Section 8.2 above –where the vehicle passes its generated data to the proprietary 'cloud' of the vehicle manufacturer who will manipulate and process the data to sell as value added services to the car owner, and, aggregated, sell to other third parties.

However, there is a recognition that some data must be made available and free, in a cybersecure environment. Automotive manufacturers generally refer to this as 'safety' data, but it is broader than this. As far as the European Commission is concerned, they have also to maintain the 'four pillars' of the EU, including an open and fair market.

8.11   Mixed AV and Conventional Traffic

For the foreseeable future, automated vehicles will share the roads with non-equipped entities, and will have to be operable within safe and feasible contexts.
In theory, the reaction time of an automated vehicle can be significantly smaller than that of a normal driver, leading to a significant increase in road capacity.

However, liability issues on the responsibility of automated vehicles are likely to force vehicle manufacturers to design their vehicles to be fairly conservative, a serious problem when mixed vehicles (automated and conventional) will be on the roads together.

How will vehicles with speed management systems interact with unequipped vehicles? Will the unequipped vehicles travel faster and continually overtake so that the speed management system gets finally switched off by a dissatisfied driver/user? Or if he cannot do that, or it is an automated vehicle, will its journey continually be pushed back, and will the accident risk increase sufficiently because of imprudent overtakes by frustrated unequipped drivers, that the automated vehicle will slow down further and further because of the additional risk. How is that going to work out?

As stated in earlier chapters, AVs may not be safer than an average driver and may increase total crashes in the mixed traffic period Schoettle and Sivak (2015).

It has been found that AVs were involved in more crashes per million miles travelled than conventional vehicles, although they were often/usually not at fault. The consequences have to be that AVs will tend to perform more cautiously compared to human-driven vehicles for safety and liability reasons, and this may tempt human driven vehicles to adopt risky behaviours thus introducing new risks.

As J3016 points out that, a Level 3–5 ADS feature that operates a vehicle on open roads in mixed traffic, and does so in environments that include inclement weather, faces a significantly higher technological bar in terms of ADS capability by virtue of the more complex and unstructured operational design domain.

Leaving generous margins to allow for the weaknesses of human drivers, or stopping, thus reducing the available space for other vehicles, will reduce the capacity of the existing road network for many years to come. However, as fully automated vehicles will not abound for a decade or so, road authorities have time to plan, and governments time to invest in additional road infrastructure. But the evidence is clear, AVs will increase the requirement to expand the road system, and road operators and governments need to plan for this now, rather than when it is too late.

Conventional drivers can and will adapt to the presence of AVs on the roads, which will be a gradual infiltration, and will adjust their driving behaviour to cope. Regulators would be wise to stiffen the penalties for lane jumping, especially where there is inadequate headway, and should enforce the regime.

8.12   AV Acceptability

Acceptability: European drivers can't be persuaded to hand over the simple changing of gears to automation, and most cars sold in Europe are still manual gear shift. Opinion is widespread that that AVs would not drive as well as human drivers and that users would feel uncomfortable travelling in an AV, and this has been backed up by other research. Users' acceptance towards AVs will of course grow as they become more familiar on the roads, but initial fears will inhibit early take up and affect early business models.

Many of the forecasts, particularly by government and automotive makers, assume that car owners and transport users will 'rush' to adopt the new AV technologies and MaaS. It is likely that the unsatiated demand groups referred to in Section 8.9 will be very keen to adopt the new technology so long as it is affordable, but, as in the statement above, it is difficult to see, particularly European, drivers rushing to the technology.

Yap et al. (2016) found that the joy of driving is of relevance for users choice of travel mode, which is reflected in users favouring the choice of the manually driven over the AV. Contrary to what was initially expected, attitudes regarding service reliability and work productivity had a secondary role in the total utility, indicating that the potential advantages of using an AV are not perceived by today's travellers.

A study on the quality of transport European Commission (2014) found that 41% of respondents would not be willing to connect their vehicles, although this survey is now quite dated and attitudes might be changing, however, the analysis is interesting.

The UK Department for Transport (2019) indicate that research has suggested that the potential benefits of self-driving vehicles are not well understood, and shared services hold limited appeal. For example, 49% of respondents to the Transport and Technology Public Attitudes Tracker were unable to name any advantages of self-driving vehicles.

Governments, planners, automotive manufacturers, futurologists in general, have to stop getting carried away with their dreams. For the foreseeable future a substantial proportion of drivers will not elect to use AVs, and some will not want connected vehicles.

Until there are a majority of AVs on the roads and governments can prove them to be safer – probably by the late twenty-first century – by which time the whole transportation paradigm may have changed – will government be able to 'prove' that manually driven cars are an unacceptable danger, and protest groups and doctors demand their removal from the roads.

In the meantime, get used to a mixed-driving paradigm.

8.13   Low Latency Communication

In an environment where the communication exchange must often have low latency – i.e. have very rapid response, an agreement on a low latency communication channel is urgently overdue.

As acknowledged by the C-ITS platform report (2016), users are not concerned about which communication technology is used to transmit C-ITS messages, but will more and more expect to receive all traffic and safety information seamlessly across Europe. (Similar research indicates similar responses in the United States and Australia). However, there may be resistance to use the technology if it incurs ongoing connection charges, particularly for the transfer of information that may have general safety benefit for all, but no specific identified benefit for the person paying the charges. Also, we may assume that some who do not pay will be cut out of critical safety services support.

Until 2016, 5.9 GHz DSRC (so called G5) was seen as the prime means of providing C-ITS communications, but the ISO 21217 C-ITS architecture (and its predecessors) have always been supporting multi-media interfaces, and because the allocated bandwidth is limited at 5.9 GHz (more so in Europe than the United States) it was recognised that the 5.9 GHz communication should be reserved for low latency safety applications, with other C-ITS applications using the by then becoming widespread 4G communications, or in some cases the already widespread 3G communications.

The powerful advantages of 5.9 GHz DSRC as specified in a number of standards, but primarily based on IEEE 802.11 (WiFi/wireless LAN communications standard, variant '802.11P' [otherwise called WAVE]) (called G5 in Europe) are short latency, limited interference, low sensitivity to weather conditions and the fact that it does not require subscription to a mobile operator's network. Importantly, it operates on an ITS station to ITS-station basis, within a range of more than 500 m, so does not need cell-based network coverage.

In respect of support of cellular technologies, there is a very high 3G mobile network coverage in most developed countries, and a growing coverage of 4G packet switched networks, forecast to equal that of 3G in populated areas by 2020, and a significant part of the population equipped with a mobile phone with data transfer capabilities. A high proportion of new vehicles are equipped with 4G communications. 3G can achieve speeds of up to 14 mbps (though usually much worse), and fourth-generation (4G) Long-Term Evolution (LTE) can offer 150 mbps, if conditions are good. But a major challenge is that some 3G networks are being refarmed to support 4G.

In 2016 3GPP, cellular phone consortium, announced the advent of 5G, to be specified by 2019. 5G is packet switched, like 4G, and supports IMS/IP protocols, but is multi-layered, offering much higher capacity – but cell coverage is much smaller and requires a much greater deployment of cells. Commercial viability in rural low population areas is still questionable. In any event, there are so many cells required for ubiquitous coverage that it will at best take more years than 4G deployment has taken to date, which puts widespread coverage well into the 2020s. (British Telecom, for example, are promoting expansion of 4G coverage to rural areas of UK by 2024, and say that 5G coverage will take longer, 'if at all viable' for rural areas).

But, in the long term, 5G is seen as a long-term support to, possibly alternative for, IEEE 8011.2 P WAVE/G5, but although there is limited city centric rollout commencing 2019, it is expected that widescale coverage will not be prevalent before 2024–2026.

While 5G is seen as a powerful carrier in the long term, several organisations are promoting its use for C-ITS in the near term, using a patented system called C-V2X.

In 3GPP Release 15, C-V2X includes support of both direct communication between vehicles (vehicle-to-vehicle [V2V]) and traditional cellular-network based communication. Also, C-V2X provides a migration path to 5G-based systems and services, which implies incompatibility and higher costs compared to 4G-based solutions. It is being touted as a migration path to 5G C-V2X, but will not be compatible with the 5G version, so claims being made for it regarding results of current 4G C-V2X trials, are not relevant for its 5G descendant, even if they use the same principles.

In C-V2X, the direct communication between vehicle and other devices (V2V, vehicle-to-infrastructure [V2I]) uses an interface called PC5. PC5 is a reference point where the user equipment (UE) directly communicates with another UE over the direct channel. In this case, the communication with the base station is not required – similar to 802.11P WAVE, but a different methodology. In system architectural level, its proximity service (called ProSe) is a feature that specifies the architecture of the direct communication between UEs (user equipment). In 3GPP RAN specifications, 'sidelink' is the terminology used to refer to the direct communication over PC5.

Note: PC5 interface was originally defined to address the needs of mission-critical communication for public safety community (Public Safety-LTE, or PS-LTE) in release 13. In release 14 onwards, the use of PC5 interface has been expanded to meet various market needs, such as communication involving wearable devices such as smartwatch.

In C-V2X, PC5 interface is re-applied to the direct communication in V2V and V2I.

In addition to the direct communication over PC5, C-V2X also allows the C-V2X device to use the cellular network connection in the traditional manner over Uu interface. Uu refers to the logical interface between the UE (user equipment) and the base station. This is generally referred to as vehicle-to-network (V2N). V2N is a unique use case to C-V2X and does not exist in 802.11p based V2X given that the latter supports direct communication only. However, similar to WLAN-based V2X also in case of C-V2X, two communication radios are required to be able to communicate simultaneously via a PC5 interface with nearby stations and via the UU interface with the network.

So the claimed advantage of C-V2X is that it integrates both direct communication and cellular communication (which is achieved in the traditional ISO 21217 by functional separation of applications from the communications carrier technology).

The C-V2X mode 4 communication relies on a distributed resource allocation scheme, namely sensing-based semipersistent scheduling, which schedules radio resources in a stand-alone fashion in each user equipment. The technology attempts to competitively assign each one a periodic 10 ms block (the RB). The RBs are of fixed size while the transmissions are not, which is spectrum use inefficient and there are a considerable number of parameters that have to be 'tuned' to make the system work. Proponents claim that in the 5G paradigm, this will work faster and more efficiently, and without the bandwidth limitations of 802.11P-WAvE/G5, but assumes that 5G has the whole of the 5 GHz spectrum available to it, and as it appears that Mode 4 C-V2X cannot co-exist with G5, this is a major problem.

While 3GPP defines the data transport features that enable V2X, it does not include V2X semantic content but proposes usage of ITS-G5 standards like CAM, DENM, BSM, etc., over 3GPP V2X data transport features.

Proponents of C-V2X (primarily its patent holders, some Telcos and a couple of automotive manufacturers, known as the 5GAA Alliance) propose that deployment of 802.11P-WAVE technology is deferred until 5G C-V2X has the opportunity to prove its superiority. The weaknesses of this argument are:

a) Roll-out beyond city centres of 5G will take till the mid-2020s (at best)
b) (and far more significantly) it is based on the assumption that C-V2X can operate using the whole 5 GHz spectrum, whereas 802.11P-WAVE/G5 is placed at the middle of that spectrum, and, to meet radio regulator requirements, a new use of the spectrum MUST prove that it can co-exist with existing uses. Radio engineers advise that C-V2X cannot co-exist with G5 without introducing latency and communication drops into G5.

Hence C-V2X advocates do not want rollout of 802.11P-WAVE/G5 before 5G is widely rolled out. It is claimed that advocate patentholders of C-V2X are much more interested in removing 802.11P-WAVE/G5 for other commercial reasons than believing that PC5 is superior to 802.11P/G5 DSRC (but, however likely, that can't be proven).

Vehicle communication technologies need to operate in a highly dynamic environment with high speed differences between transmitters and receivers and need to support extremely low latency for safety-critical applications, among other requirements. Aspects such as security and robustness are critical in vehicle and infrastructure communications. 802.11P-WAVE + IEEE 1609/ISO 21217 and associated standards, and ISO 21177 secure communications, have proved that they can achieve this with low latency.

The decision by Volkswagen (see Chapter 3) and European Commission (2019/ October) shows a clear preference and commitment to IEEE 802.11P-WAVE/G5 for near-term solutions (Level 1 and 2) by a significant number of important actors – at least on the European scene.

China, the world's largest vehicle market, is pressing ahead with implementing 5G solutions.

In the long term, 5G will play an important role, but the future of contumelious solutions such as C-V2X PC5 is very dubious, because it cannot co-exist with existing uses at 5.9 GHz.

8.14 Roads Could Be Allocated Exclusively to AVs

It has been proposed that some roads could be allocated exclusively to AVs, or special roads (with very high standards of road marking), built for them. Other ideas being discussed are that some HOV (high-occupancy vehicle) lanes could be reassigned for exclusive use of AVs, that motorways could have one lane allocated to automated and 'connected' vehicles, etc.

The current road network is designed to accommodate the existing manually driven transport. AV developments are currently focussed to operate in an integrated manner with existing transport solutions in this current paradigm. And it is essential that they can do this. But is that the best way to gain the benefits of AVs? It may be expected that AVs (and possibly connected vehicles) will move more efficiently, closer together and without the same impact on the road surface where they are segregated from manually driven vehicles – this could be on roads dedicated to AVs, or lanes dedicated to AVs (much as there are currently HOV) lanes in many parts of the world. This could also assist AVs to select the best route and link in with the local plans for traffic movement.

This will have implications for all aspects of the infrastructure, including the layout, the road surface and signalling. AV only lanes/roads require markings, signals and signs to be maintained to a higher standard than at present to make sure the instructions can be followed. However, selecting some lanes/roads as AV only (or building new ones instead of general road expansion to handle rising demand) could be a faster and lower cost option than upgrading markings and signals and signs on *all* roads, – and there will be pressure to do this to enable AVs to operate safely.

Also, where platooning is envisaged, there is a greater need to maintain road surfaces as AVs may be less able to adapt to potholes, a particular risk with platoons where vehicles travel much closer to each other.

In order for the full benefits that connected and autonomous vehicles (CAVs) could offer to be achieved, detailed modelling and analysis will be the key to making the correct decisions.

Discussions with local administrations, who, in most nations, have been under administering under increasing financial pressure, especially since the 2008 banking problems, produce the response that 'it would be very nice, but we have no resources to even consider this'.

The pressure on local administrations is understandable. However, it is a defensive response to avoid facing difficult issues that they face all the time in other contexts. There is no technical difference to make a bypass for automated vehicles than to make a bypass for all, or conventional, vehicles. There is no technical difference to provide an extra lane on a motorway for AVs only than providing an extra lane for all vehicles. There is no technical difference to reassign an HOV lane to an AV lane, etc. These issues are political, and about social acceptance.

Yet even in this time of austerity we have seen significant diversion of investment, often to the detriment of increasing congestion (at a time when rising congestion and its associated pollution are paramount issues for citizens) to assigning lanes to buses, or cyclists, or building cycling superhighways, because it fits with political objectives. If and when the time comes that there is political justification for roads to be allocated exclusively to AVs or connected vehicles with some AV capabilities, ways can be found. The issues, however, are not technical, an only to a limited extent, financial. But as the realisation of AVs on the roads is still a future dream. At the moment, policy change is not yet ready.

One facet will help planners who are arguing the case for separated highways/lanes for AVs, and that is public concerns/fears about the safety of AVs. As stated in a previous chapter, research Schoettle and Sivak (2014) has indicated that there was likely to be a low level of trust on the safety of AVs, and concluded that opinion was widespread that that AVs would not drive as well as human drivers and that users would feel uncomfortable travelling in an AV, and this has been backed up by other research. Planners can offer separated lanes/highways as a way of alleviating such fears.

In the context of building the federal highway system in the 1950–1980s period in the United States, or 1960–2000 period of the expansion of autoroutes/autobahns/ motorways in Europe, and continuing town/city rocades, peripheriques, bypasses and ring roads since the end of WW2, these issues are relatively minor, and because of the improved safety of life cost benefit of AV only roads/lanes, will be relatively easy to cost justify.

But so long as the AV is a thing of the future, and not realised on the street, the political will not be there.

8.15 Automated and Connected Vehicles Bring New Requirements

Of course, automated and connected vehicles bring new requirements and challenges to the regulators in respect of traffic law, liability, security, access to data, protection of personal data, particularly for Level 4 and 5 vehicles.

As discussed in the last two sections, AVs bring new requirements which will require changes in respect of traffic law, and local bylaws. Chapter 8.17 discusses the delays in making such changes.

The prime instrument of the UN Economic Commission for Europe (UNECE) in respect of road transport is the '1968 Vienna Convention on Road Traffic', which is an accord among participating members of UNECE. The convention covers road traffic safety regulations and as such establishes principles to govern traffic laws.

One of the fundamental principles of the convention is the concept that *a driver is always fully in control and responsible for the behaviour of a vehicle in traffic.* And this is enshrined in the laws/traffic regulations of most countries around the world.

The advent of the AV changes that paradigm, and this instrument has to be updated by UNECE, then adopted by countries around the world, before Level 4 and 5 AVs can be legally operated on their roads.

New amendments to UNECE recommended regulations have already been adopted and came into force in March 2016. The key amendment allows a vehicle to drive itself, so long as the system 'can be overridden or switched off by the driver'. But that implies that a driver must be present and able to take the wheel at any time. This interpretation has to be implemented in national regulations, and countries at the leading edge of AV developments are in the process of doing this in order to enable Level 3 – conditional automated driving. Netherlands and Germany already have this in place. UK and France are not far behind.

As an example, in the UK, the Automated and Electric Vehicles (AEV) Act, was passed in July 2018. Implementation is anticipated via a number of statutory instruments within the next few years.

This act requires the secretary of state to maintain a list of motor vehicles that are:

> 'designed or adapted' to 'be capable of safely driving themselves'; and
> may lawfully be used when driving themselves on roads or other public places in Great Britain.

But it is not yet clear who will bear the expense of collating and maintaining this list. Once the list is published, the secretary of state will have two years to report to parliament as to the effectiveness of that listing.

In order to enable Level 4 and 5 driving, UNECE have to make further revisions, both at the WP1 level and at WP29 level, and additional changes also to UN R79 (Steering), and UN R48 (Lighting).

Amongst the new requirement, AVs, and indeed all connected vehicles need to exchange data in what ISO 21217 calls a 'Bounded Secure Security Domain,' i.e. need cybersecurity. This is of such importance that the next section of this chapter addresses these issues.

8.16 Cybersecurity

Cybersecurity is another major issue. In part for privacy, but more particularly to prevent hacking into vehicle control systems, or indeed the ADS itself.

Cybersecurity first cropped up in Section 8.2.

- the EC and US DoT will have to regulate regarding
  a) *Data ownership*

b) *Data required for safety services*
c) *Data access*
d) *Cybersecurity*

..........*The data exchanged must be freely available within cybersecure transactions, and must be provided with very low latency.* As part of that process, cybersecurity standards need to be agreed upon to define the minimum security embedded in the hardware; as well as what the boundaries are for software and connectivity.

In this context, aspects such as data sharing, security and privacy are of paramount importance. Cybersecurity is a prerequisite for data exchange between connected vehicles because of the need to protect networks, computers, programs and data from attack, damage or unauthorised access.

The connected vehicle and connected infrastructure requires available data transmission frequencies, low-latency, trusted, secure and fail-safe data transmission protocols and harmonised data syntax that ensures safe interoperability OECD/ITF (2015).

Tillemann and McCormick (2016) postulated that sharing AV crash and incidents data would contribute to the improvement of AV technologies, learning from real accident data in order to make AVs safer. Clearly such data exchange can only be made in a cybersecure and anonymous way.

The Network and Information Security Directive (NIS) European Commission (2015b) will have an impact on cloud services that may be associated with connected vehicle. It remains to be decided if additional regulation in this regard is needed.

From an international perspective, the 'Harmonization Task Group' (HTG) (a collaboration of US DOT, the EU JRC, EC DG CNECT and the Transport Certification Authority [TCA] of Australia), have been working to harmonise standards and jointly address issues facing connected vehicles, and they have produced a set of functional specification and recommendations for cybersecurity of C-ITS European Commission (2015c).

In the United States, National Highway Traffic Safety Administration (NHTSA) policy guidance document (2016), cybersecurity is covered by recommending that manufacturers and other entities to follow a robust product development process based on a systems-engineering approach, including systematic and ongoing safety risk assessment for the AV system, the overall vehicle design into which it is being integrated and, when applicable, the broader transport ecosystem.

The Alliance of Automobile Manufacturers (AAM) formed a voluntary Information Sharing and Analysis Centre (Auto ISAC) in 2014 to target the threat of hackers (2014). This defines an ISAC, which is a trusted, sector-specific entity that can provide a 24-hour per day and 7-day per week secure operating capability that establishes the coordination, information sharing and intelligence requirements for dealing with cybersecurity incidents, threats and vulnerabilities. Adopting a vulnerability disclosure policy is also encouraged.

In May 2018 the EC JRC published its 'Certificate Policy for Deployment and Operation of European Cooperative Intelligent Transport Systems (C-ITS)' and announced that it had funding for trials and early years of operation of the system (2018).

The certificate policy defines the European C-ITS Trust model based on Public Key Infrastructure (PKI) within the scope of an overall 'EU C-ITS Credential Management System' (EU CCMS). To ensure those main objectives, a security architecture with support of a PKI using commonly changing pseudonym certificates, has been developed.

It defines legal and technical requirements for the management of public key certificates for C-ITS applications by issuing entities and their usage by end-entities in Europe. The PKI is composed at its highest level by a set of root Certification Authorities 'enabled' by the Trust List Manager (TLM), i.e. whose certificates are inserted in an European Certificate Trust List (ECTL), which is issued and published by the central entity TLM.

This policy is binding to all entities participating in the trusted C-ITS system in Europe. The policy can therefore be used as guidance to assess which level of trust can be established in the received information by any receiver of a message authenticated by an end-entity certificate of the PKI.

To allow assessment of trust in certificates, this policy defines a binding set of requirements for the operation of the central entity TLM and the definition and management of the ECTL. Consequently, this document defines the following aspects related to the ECTL:

- Identification and authentication of principals obtaining PKI roles for the TLM including statements of the privileges allocated to each role.
- Minimum requirements for the local security practices for the TLM, including physical controls, personnel controls and procedural controls.
- Minimum requirements for the technical security practices for the TLM, including computer security controls, network security controls and cryptographic module engineering controls.
- Minimum requirements for operational practices for the TLM including registration of new root certification authority certificates as well as temporary or permanent deregistration of existing included root certification authorities as well as publication and distribution of the ECTL updates.
- ECTL profile, including all mandatory and optional data fields contained in the ECTL, used cryptographic algorithms, as well as the exact ECTL format and recommendations for processing of the ECTL.
- ECTL certificate lifecycle management, including distribution of ECTL certificates, activation, expiration, and revocation.
- Management of the revocation of trust of root certification authorities when needed.

Since the trustworthiness of the ECTL does not solely depend on the ECTL itself, but to a large extent also on the root certification authorities that compose the PKI and their sub certification authorities, the policy also defines minimum requirements for certain aspects, which are mandatory for all participating certification authorities to be implemented in the certificate practice statements of these certification authorities. In particular, these aspects are:

- Identification and authentication of principals obtaining PKI roles (e.g. security officer, privacy officer, security administrator, directory administrator, and end-user) including a statement of duties, responsibilities, liabilities, and privileges associated with each role.
- Key management including acceptable and mandatory certificate signing and data signing algorithms as well as certificate validity periods.
- Minimum requirements for local security practices including physical controls, personnel controls, and procedural controls.

- Minimum requirements for technical security practices such as computer security controls, network security controls and cryptographic module engineering controls.
- Minimum requirements for operational practices of the certification authority, enrolment authority, authorisation authorities, and end-entities including topics of registration, de-registration (i.e. de-listing), revocation, key-compromise, dismissal for cause, certificate update, audit practices, and nondisclosure of privacy related information.
- Certificate and CRL profile including formats, acceptable algorithms, mandatory and optional data fields and their valid value ranges and how certificates are expected to be processed by verifiers.
- Regular monitoring reporting, alerting and restore duties of the C-ITS trust model entities in order to establish a secure operation including cases of misbehaviour.

In addition to these minimum requirements the entities running the root certification authorities and sub certification authorities may define their own additional requirements and define them in the respective certificate practice statement.

The certificate policy explicitly identifies the applicable European regulations to which the TLM shall conform, including data protection, privacy, access to information and lawful interception legislation.

This has been followed in 2019 by the publication of ISO TS 21177, which is consistent with this policy, but broader in application scope.

The scope of ISO 21177, *Intelligent transport systems – ITS-station security services for secure session establishment and authentication between trusted devices* (n.d.) is to provide:

> "specifications for a set of ITS station security services required to ensure the authenticity of the source and integrity of information exchanged between trusted entities:
>
> - devices operated as bounded secured managed entities, i.e. 'ITS Station Communication Units' (ITS-SCU) and 'ITS station units' (ITS-SU) specified in ISO 21217, and
> - between ITS-SUs (composed of one or several ITS-SCUs) and external trusted entities such as sensor and control networks.
>
> *These services include authentication and secure session establishment, which are required to exchange information in a trusted and secure manner.*
> *These services are essential for many ITS applications and services including time-critical safety applications, automated driving, remote management of ITS stations (ISO 24102-2), and roadside/infrastructure related services."*

A cybersecurity framework therefore already exist that meets European requirements for cybersecurity and operates within GDPR data protection and privacy requirements.

Whether North America establishes a single framework, compliant to ISO 21177, or joins with the European scheme, is yet to be decided. Australia will similarly decide whether to set up a similar framework or join with the European scheme is currently under investigation within Austroads.

C-ITS service providers, and system suppliers now need to enrol and participate in the JRC managed scheme, and an equivalent needs to be established elsewhere around the world.

Problems remaining include the long-term funding of the management of cyber-security and who the long-term managing entity will be. A second problem is that vehicle manufacturers (despite incidents like *hackers-remotely-kill-jeep-highway* (Wired. 2015/07)) continue to claim that security can only be ensured through their proprietary security provisions, and that required data between vehicles can be exchanged between OEM 'clouds'. This proposition fails woefully on the grounds of the latency involved in such communications is structurally unworkable in a C-ITS safety context. It also assumes that a high-speed, high bandwidth network is available everywhere and at all times, which is clearly never going to be the case, and this must be seen in the context that OEMs have had their security hacked many times (not just the example provided above). Their case is, even by their own peers' views, and the position of UNECE WP29 GVRA, and the EC, considered a lost cause. But so long as OEMs cling to it, it delays the introduction of connected vehicles and AVs.

The decision by Volkswagen (Business Insider 2019/10) to adopt the EC and ITS sector preferred route, and the decision of others, such as Volvo to follow, at least for Level 1 and 2 automation should hopefully break the logjam and now enable rapid implementation of 'connected' vehicles, and pave the way for AVs later.

Most OEMs currently have a preference to prefer company specific proprietary cybersecurity. This approach will complicate and delay instantiation of more advanced paradigms, particularly with V2X where X is anything other than a vehicle by the same manufacturer.

8.17   Changing Speed Limits and Even Getting Signs Put Up Can Take Years

> *In practice just changing a speed limit and putting up the new signs can take 4–5 years.*

In democracies, in order to change a speed limit, even put up a road sign, in different ways in different countries, you have to work through a similar process. First of all, an application has to be made (these things don't happen by chance). The application will go through some sort of professional review. It will then most likely go through some form of public consultation. Objections will have to be heard and dealt with, dismissed or accommodated. It will often have to pass from a local level to a district level, or vice versa, and maybe there will be a consultation at that level. If it is a major change, the highway authority will also probably need to be involved. Only then can the work be physically planned and fitted into available budget, which may well already be fully committed for that financial year, deferring the work to a following year. Of course, practices vary somewhat from country to country, but the same sort of processes have to be worked through – and regardless of the variations, the one thing you can be sure of about the process is that 'it ain't gonna be quick'.

Those planning changes for the AV paradigm just have to know and understand this.

Fortunately, the unrealistic hype about the speed of introduction of AVs raises awareness that changes/adaptations will be required in good time, so planners should press ahead with getting the legal framework in place now, and work to find budget to make required changes over the next few years so that the infrastructure can seamlessly cope with the introduction of AVs when they arrive.

Entrepreneurs and venture capitalists need to be aware that this is a long-term investment that will not produce rapid return on investment.

Government authorities and regional authorities (such as the EU/EC) who see the societal benefits in the long term by reducing road deaths and injuries, and possibly reducing pollution and congestion, need to take heed that while in societal terms there is a great return on investment for 'connected' vehicles, and in respect of automated vehicles and MaaS, in the commercial world, the business case is much weaker, and the benefits do not necessarily rest with the investor, only indirectly through improving travel safety and the travel experience, and should consider what incentives they can provide to gain these societal benefits. Governments have shown, for example, by providing tax relief and cash incentives, they can accelerate the take-up of electric powered vehicles, despite the operational disadvantages and cost of these vehicles. Incentives for 'connected' vehicles could, for example, be a reduced taxation rate (either on purchase or annual taxes), a one-off cash reward for purchasing and driving a 'connected' vehicle, or by granting different access rights into controlled zones, etc. (incentives for MaaS are considered in Chapter 9).

8.18 Political Decisions Needed

Political decisions need to be made as to whether the provision of data and other C-ITS service provision to vehicles from the infrastructure should be a local authority service, a regional service or a national service, or a commercial service, or a combination of all four.

It is likely that different nation states will implement service provision in different ways. We have seen in Chapter 4 that the Czech Republic are looking to implement a single national traffic information centre; Sweden has already done so. While it is likely that decentralised Nation States (e.g. Germany) will work on a different model.

The decision therefore has to be made at an early stage by the national government, otherwise there is the risk that a forward thinking local or regional authority might start to implement to a different model, making a consistent national policy difficult to implement. As some of the larger cities in Europe at least (but also at least New York in the United States) are pressing ahead with dealing with these issues on a local level, central government need to be advising on what the national model will be, before these pace setters are too far down the line.

Venture capitalists and investors need to take central government positions, and indeed the lack of progress of a government policy in these areas, into account in their risk assessments.

8.19   Role of Government

The role of governments – central and local – in the evolution of automated vehicles/MaaS will be very significant in determining the progress, success (or lack thereof) of the automated driving/Maas paradigms.

Here the potential solution lies in the issue raised as a potential problem. Section 8.17 dealt specifically with data access and travel control points. AVs and therefore MaaS involving AVs will not happen until governments determine their policies, adjust their regulations and set their policies (such as those dealt with in Section 8.18).

National government policies, while focussed on national requirements, are heavily influenced by the UN ECE, which provides internationally agreed by consensus 'regulations' for nations to consider to adopt for road safety and vehicle regulations.

UNECE WP.29 prepares the work of the UNECE Inland Transport Committee to develop and adopt harmonised vehicle regulations. UNECE WP29, 'World Forum for Harmonization of Vehicle Regulations' manage three UN Agreements:

- UN Regulations (1958 Agreement)
- Technical Inspections (1997 Agreement)
- Global Regulations (1998 Agreement)

which provide the legal framework allowing contracting parties (member countries) attending the WP.29 sessions to establish regulatory instruments concerning motor vehicles and motor vehicle equipment.

Overall, the regulatory framework developed by the World Forum WP.29 allows the market introduction of innovative vehicle technologies, while continuously improving global vehicle safety. The framework is designed to enable decreasing environmental pollution and energy consumption, as well as the improvement of anti-theft capabilities.

UN regulations, annexed to the 1958 agreement comprise:

- United Nations Global Technical Regulations (UN GTRs), associated with the 1998 agreement;
- UN Rules, annexed to the 1997 agreement

UN regulations contain provisions (for vehicles, their systems, parts and equipment) related to safety and environmental aspects. They include performance-oriented test requirements, as well as administrative procedures. They evaluate the performance of vehicle components and subsystems. They operate within a 'type-approval' paradigm, where a vehicle is type-approved, i.e. is certified that it meets the type approval requirements, before it is allowed to be sold and used on the roads. Type approval (of vehicle systems, parts and equipment), is designed to assure the conformity of production (i.e. the means to prove the ability, for manufacturers, to produce a series of products that exactly match the type approval specifications) and the mutual recognition of the type approvals granted by Contracting Parties.

UN GTRs contain globally harmonised performance-related requirements and test procedures. They provide a predictable regulatory framework for the global automotive industry, consumers and their associations. They do not contain administrative provisions for type approvals and their mutual recognition.

UN Rules concern periodical technical inspections of vehicles in use. Contracting Parties reciprocally recognise (with certain conditions) the international inspection certificates granted according to the UN Rules.

This regulatory framework is also designed to foster and facilitate cross-border trade, since provisions established under the 1958 agreement include the reciprocal acceptance of approvals of vehicle systems, parts and equipment issued by other Contracting Parties.

UNECE Regulations are offered to participating countries who make national decisions whether or not to adopt a UNECE regulation. (The European Commission takes a view for the EU member states and tries to obtain consensus of most if not all, towards new UNECE vehicle regulations. However, adoption remains an issue of 'subsidiarity', and is a Nation State decision). Once a Nation State adopts a UNECE vehicle regulation, it becomes a 'Contracting Party', and 'Contracting Parties' maintain, revise and evolve existing regulations (through WP29).

Where GTR and UN Regulations treat the same subject, UN Regulations are required to conform with the provisions of their respective GTR; on the other hand, 'UN Rules' concern the harmonisation of vehicle inspection requirements to facilitate international road traffic.

WP29 comprises the following major subcommittees:

- Working Party on Noise and Tyres (GRBP) (former GRB)
- Working Party on Lighting and Light-Signalling (GRE)
- Working Party on Pollution and Energy (GRPE)
- Working Party on General Safety Provisions (GRSG)
- Working Party on Passive Safety (GRSP)
- Working Party on Automated/Autonomous and Connected Vehicles (GRVA)
- And a dormant group – Former Working Party on Brakes and Running Gear (GRRF)

WP.29 has created the committee called GVRA (General Requirements for Vehicle Automation) specifically to advance the regulatory framework for AVs and connected vehicles. It is progressively placed as a priority topic on the WP.29 agenda.

The Working Party on Automated and Connected Vehicles (GRVA) was created, along with subworking groups, in order to address various automated driving-related regulatory issues. Recently, a 'framework document on automated vehicles' (2019) was developed by GRVA and adopted by WP.29 to provide guidance for the subworking groups by identifying principles to facilitate and guide discussions and activities on AV performance. One key subgroup of the GRVA identified in the framework is Functional Requirements for Automated Vehicles (FRAV). Key work priorities for FRAV currently include functional requirements for the combination of different functions for driving, e.g. longitudinal control, lateral control and environment monitoring. GRVA priorities are

a) Functional requirements
b) New assessment/test methods

c) Cyber security and software updates
d) Data storage system for automated driving (DSSAD)

and targets a milestone date of March 2020 for delivery of first considerations on these subjects to be produced.

In March 2019, GVRA submitted to WP29 the first version of its 'framework document on automated/autonomous vehicles' (2019) in which it identifies key areas to be defined in UNECE regulations:

a) **Common Principles:** a list of common principles with brief descriptions and explanation. It is expected these would form the basis for further development within the GRs.
b) **Operational Domain (OD) (automated mode):** For the assessment of the vehicle safety, the vehicle manufacturers should document the OD available on their vehicles and the functionality of the vehicle within the prescribed OD. The OD should describe the specific conditions under which the automated vehicle is intended to drive in the automated mode. The OD should include the following information at a minimum: roadway types; geographic area; speed range; environmental conditions (weather as well as day/night time); and other domain constraints.
c) **System Safety:** When in the automated mode (OD), the automated vehicle should be free of unreasonable safety risks to the driver and other road users and ensure compliance with road traffic regulations.
d) **Failsafe Response:** The automated vehicles should be able to detect when a problem is encountered or when the conditions for the OD are not met anymore. In such a case the vehicle should be able to transition automatically (minimum risk manoeuvre) to a minimal risk condition with or without take over request.
e) **Human Machine Interface (HMI)/Operator information:** Automated vehicle should include driver engagement monitoring in cases where drivers could be involved (e.g. take over requests) in the driving task to assess driver awareness and readiness to perform the full driving task. In addition, automated vehicle should allow interaction with other road users (e.g. by means of external HMI on operational status of the vehicle).
f) **Object Event Detection and Response (OEDR):** The automated vehicles shall be able to detect and respond to object/events that may be reasonably expected in the OD.
g) **Validation for System Safety:** Vehicle manufacturers should demonstrate a robust design and validation process based on a systems-engineering approach with the goal of designing ADSs free of unreasonable safety risks and ensuring compliance with road traffic regulations and the principles listed in this document. Design and validation methods should include a hazard analysis and safety risk assessment for ADS, for the overall vehicle design into which it is being integrated and when applicable, for the broader transportation ecosystem. Design and validation methods should demonstrate the behavioural competencies an ADS would be expected to perform during a normal operation, the performance during crash avoidance situations and the performances during a crash. Test approaches may include a combination of simulation, test track and on road testing.
h) **Cybersecurity:** The automated vehicle should be protected against cyber-attacks in accordance with established best practices for cyber vehicle physical systems.

i) **Software Updates**: Vehicle manufacturers should ensure system updates occur as needed in a safe way and provide for after-market repairs and modifications as needed.

j) **Event Data Recorder**: (Description: TBD)

k) **Data Storage System for Automated Driving vehicles (DSSAD)**: The automated vehicles should have the function that collects and records the necessary data related to the system status, occurrence of malfunctions, degradations or failures in a way that can be used to establish the cause of any crash.

While it is clear that the priority for work undertaken in GVRA is near-term systems likely to be implemented in the near future, of concern is that its ambition for a framework document on automated/autonomous vehicles, is very short-sighted. For example:

- Operational Domain – This is necessary information, but the way it is written implies that AVs only work some of the time. These are limitations for J3016 Level 1–3 automation, and what J3016 describes as 'driving automation systems' (DAS), and specifically segregates these levels from what it describes as 'automated driving system' (ADS). In the context of this book we are looking at barriers to ADS, not DAS (see Chapter 2).
- Failsafe Response: '..... In such a case the vehicle should be able to transition automatically (minimum risk manoeuvre) to a minimal risk condition with or without take over request' implies that there is always a driver in the vehicle.
- Human Machine Interface (HMI)/Operator information: 'Automated vehicle should include driver engagement monitoring in cases where drivers could be involved (e.g. take over requests)' .....again implies that there is always a driver in the vehicle.

These DAS conditions of course need to be provided for in UN Regulations (and quickly) because these features are already appearing in vehicles, but the title of GVRA – *general requirements for vehicle automation*, and its claimed scope is about vehicle automation, not just driver assistance.

'General requirements for automated vehicles and automated driving' is clearly a high urgency need that is not yet satiated, nor even addressed. This needs to be undertaken quickly.

UNECE also has WP1, which sits as a Global Forum for Road Traffic Safety and remains the only permanent body in the United Nations system that focuses on improving road safety. Its primary function is to serve as guardian of the United Nations legal instruments aimed at harmonising traffic rules. The Conventions on Road Traffic and on Roads Signs and Signals of 1968, and other UNECE legal instruments that address the main factors of road accidents (road user behaviour, vehicle and infrastructure) are tangible contributors to improved road safety. Many countries across the world have become Contracting Parties to these legal instruments and benefit from their implementation. These Contracting Parties are also the key driving forces keeping these international road safety conventions up-to-date.

The Global Forum for Road Traffic Safety (WP.1) focuses on improving road safety through the harmonisation of traffic rules. The forum oversees the application of the 1949 and 1968 Conventions on Road Traffic and the 1949 and 1968 Conventions on Road Signs and Signals. There has been a growing interest within WP.1 to address concerns related to automated driving, such as ensuring that existing international law is compatible with automated vehicles (AVs). The Informal Group of Experts on Automated

Driving (IGEAD) was created to assist WP.1 with its work on automated driving. In September 2018, through the work of IGEAD, WP.1 adopted a resolution on the deployment of highly and fully automated vehicles in road traffic. WP.1's key work priorities in this area relate to the need to provide guidance to support the safe deployment of AVs in road traffic. Additional work is ongoing within IGEAD to develop other guidance to determine how best to apply the principles of the 1949 and 1968 conventions in the context of automated driving, e.g. remote-controlled driving and activities other than driving for drivers of highly automated vehicles.

Clearly, for the advent of AVs, WP1 and WP29 have to cooperate and agree who does what. The document 'Collaboration and common approaches between WP.1-WP.29 on automated vehicles' (n.d.) provides the methodology for their cooperation.

At its 81st session in February 2019, the Inland Transport Committee invited WP.1 and WP.29 to continue their close cooperation to facilitate the safe deployment of automated vehicles. Given the importance of automation to the work of both WP.1 and WP.29, it is important that both parties collaborate to ensure that the evolution of international conventions on road safety is carried out in concert with safety requirements to avoid any incompatibilities. Both parties have agreed to create a joint work plan. This collaboration is intended to facilitate joint work around the Society of Automotive Engineers (SAE) Level 3 to 5 automated vehicles and their safe deployment in traffic environment.

But this work is only just commencing at the time of writing this book. However, the timing for the implementation of standards is crucial as regulating too late will result in different standards. Fortunately, as we have established in the chapters to date, there are some years before Level 4 and 5 will be ready to drive (safely) on the roads, and there is hope that the regulatory framework will by then exist to enable national regulators to adapt their regulations in a common way to accommodate this. But UN committees are not the fastest, and even when they produce their regulatory revisions, there is then the process of national adoption, so the pressure needs to be kept onto this joint work to deliver.

Investors and venture capitalists should be wary where no clear central government policy is emerging/being worked on. Venture capitalists and investors need to take central government positions, and indeed the lack of progress of a government policy in these areas into account in their risk assessments.

8.20 Fallback to Driver

Where the automated system cannot reconcile its situation, in J3016 paradigms 1–3, control is passed back to the driver. Further research is needed the time needed for the driver to takeback control. What happens if the occupants of the vehicle are asleep, have been drinking, or are not qualified drivers? These are unsolved problems to this fall-back paradigm.

As we have seen in Section 8.19, all of the early levels of automation – as J3016 calls it 'driving automation system' (DAS) – and in the systems that WP29 is currently paying attention to in its GVRA subcommittee (which cover similar levels of automation), there is the possibility that the automation system encounters something it cannot deal with, so it has to have a process to hand back control to the driver. The details of how it alerts the driver to take back control, and the steps the driver has to make to take back control, are important, but not the focus of this section.

Most of us are used to cruise control, and any touch of the brake or accelerator returns the vehicle to the control of the driver. But in these circumstances the driver has remained involved, because s(he) is still maintaining the task of steering the vehicle, so hand-back of control to the driver is (i) instigated by the driver who is (ii) still involved in the driving task – so the handover is effectively instantaneous.

Clarifying the liability circumstances in both a DAS or ADS context is crucial. At present (DAS environment) the driver is expected to remain in control of the vehicle at all times and it is clear that the driver is liable should a crash occur. However, based on the information below, that may be an unreasonable responsibility, and needs to be reviewed by regulators.

So long as the driver has a genuine opportunity to take control over a DAS car and avoid a crash (a minimum of eight to nine seconds is suggested by Kühn (2016) and others believe even this to be inadequate [more on this below]), the liability is likely to remain with the driver. But, due to the time it actually takes to recover control of the vehicle, this is probably unreasonable.

Where there is a malfunction of an DAS-automated vehicle it is important but more difficult to know who is liable in case of a collision: the manufacturer or the driver? These issues particularly concern insurers.

Stephen Casner, a research psychologist in the Human Systems Integration Division at NASA et al. (2016) writing for the Association for Computing Machinery (ACM) opines that: *' . . . . the transition will be difficult, especially during the period when the automation is both incomplete and imperfect, requiring the human driver to maintain oversight and sometimes intervene and take closer control.'*

There is much research in this area, and research continues. Perhaps this study by Casner and colleagues provides one of the most comprehensive and incisive research work in an area that is critical for DAS automation. As the consequences are so critical, and so poorly understood by much of the ITS and regulator sectors, this section looks at their findings in some detail.

Referencing more than 40 respected research studies, Casner and colleagues reviewed two kinds of emerging car automation, They drew from extensive research on the state of the art in driving together with decades of previous work that examined the safety effects of automation as it was gradually introduced in the airline cockpit and found that some problems seem counterintuitive and some paradoxical, with few of them lending themselves to simple solutions.

They first studied 'Provide advice but leave the driver in charge' systems, particularly with reference to distraction, but, importantly, also concluded that when a system (example: navigation) performs well over extended periods, drivers

' . . . .may no longer feel they need to pay close attention. Indeed, many psychological studies show people have trouble focusing their attention when there is little or nothing to attend to. In such situations, they tend to reduce their active involvement and simply obey the automation.' – even if that instruction is obviously wrong (they cite an example of a driver following a sat-nav instruction to drive over the side of a cliff [because the sat-nav mistook a footpath for a road], and the example of a crew of a Boeing 757 which flew into a mountain near Buga, Colombia, after following the directions given by their erroneously programmed flight-management system).

They determined evidence of what they call 'Skill atrophy', stating that '….. There is good evidence that cognitive skills erode when not practiced regularly'.

They observed that '….drivers got lost before the introduction of navigational systems, but they seldom led to safety-critical incidents. GPS navigation has introduced many human factors complications we did not anticipate.'

They also conclude that advisory and warning systems can cause problems for the primary task of paying attention. They reference Wiener (1985) who called this effect 'primary-secondary task inversion,' and backed it with considerable evidence from aircraft cockpits, where it is a commonplace problem.

They pay close attention to the short timeframes involved in the driving task, pointing out that an automobile can be far more dangerous than an airplane in several respects. In an airplane flying at cruising altitude of 10–12 km, the pilots might have minutes in which to respond. In a vehicle, the available time can sometimes amount to a fraction of a second, and they point out that laboratory studies of driver reactions to rear-end collision alerts show the effectiveness of these alerts falls off quickly when alert times are short.

Noting these relevant conclusions, they then turn the attention of their study to systems that directly controls all or part of an automobile.

Level 0 cars already reduce driving to a remarkably mundane task, sometimes requiring little attention from the driver and luring the driver into distraction. But they presented evidence from other studies that demonstrated that relieving drivers of even one aspect of the driving task results in reports of increased driver drowsiness and reduced vigilance when driving on open stretches of road. And they presented research that shows that the effects do not stop there.

Drivers also take more time to respond to sudden events when, for example, they use cruise control. And they conclude, importantly, that if you take drivers out of the role of active control, it is difficult to get them back in when they are needed. This has very significant impact for DAS systems. They opine that this brings real issues for Level 2 automation systems that use automation to control two or more functions of the driving task at once.

A key feature of Level 2 automation is it is generally capable of fully controlling the vehicle for limited periods in restricted situations (such as following another car during uneventful freeway cruising or during traffic jams).

J3016/NHTSA Level 2 automation assumes the human driver will continue to closely monitor the automation as it follows the car ahead. Manufacturers differ in their requirements for drivers to keep their hands on the steering wheel. Casner and colleagues observe that some systems simply require driver hands to be near the steering wheel in the case a driver takeover is required on short notice, and point out that research evidence shows that, as automation becomes more able and reliable, drivers will inevitably do things other than pay attention to driving. They point to recent driving research studying these situations, and they note that the results are not encouraging.

Casner observes that the problem of reengaging drivers to assume active control of the vehicle is quite complex with Level 2 automation. The NHSTA/UNECE (and most regulations and guidelines) require that the driver must '…..be able to determine, at any moment, what driving functions are being handled by the automation and what functions remain the responsibility of the driver.'

They point out that eye-tracking studies of airline pilots reveal they persistently mis-remember the state of the automation, even when they themselves set up the state. They rely on their memory of having pushed a button and habitually ignore system-status displays that tell the real story Sarter et al. (2007). Incidents in which pilots pressed a button to engage a speed control function only to later see their speed increase or decrease unexpectedly are commonplace.

Casner and colleagues conclude that as research shows that because such problems will be common and endemic, the only effective solution is to largely eliminate the need for attention and understanding from drivers by adding even more automation, moving to Level 3 automation, where vehicles use automation to control all aspects of the driving task for extended periods, because Level 3 automation does not require the driver's constant attention, only that the automation provides drivers a comfortable transition time when human intervention is needed.

Looking to address the problems with Level 3 automation, when drivers are needed, the system relies on what is called 'conditional driver takeover' in which drivers are summoned and asked to intervene.

Casner and his colleagues point out that the biggest challenge here that has to be overcome is that 'people have great difficulty re-establishing the driving context, or as psychologists call it "rapid onboarding". To make matters worse, automation often fails when it encounters unexpected problems, leaving the driver with only a short time to respond'.

They point out that driving researchers have begun to show drivers' onboarding times grow quickly when high levels of automation are combined with complex situations. They point out that studies of airline pilots responding to such unexpected events inspire little confidence Casner et al. (2013).

'Referring back to what they had described as "manual skill atrophy", they provide evidence that " …..Prolonged use of automation leads to deterioration of skills. Airline pilots who use high levels of automation in an airline cockpit continually complain about it". Casner et al. (2014) found cognitive skills (such as navigating and troubleshooting) were quick to deteriorate in the absence of practice.'

It is clear from evidence provided in this study, that automation systems are generally insufficiently understood and perhaps overly trusted by the user. Much of their evidence is from flight crew and crash investigations, and the recent Boeing 737800MAX crashes sadly provide further proof. There is no reason to believe that this will not be the case for AVs, indeed, because vehicle drivers are less well trained than aircraft pilots, the situation is likely to be exacerbated.

In an aircraft, moving in a controlled environment, with ground controlled separation of height, direction, and approximate speed, a pilot may be expected to monitor his automated systems. Automated vehicles manoeuvring stochastically in dense traffic, with vehicles silently communicating with each other and with the infrastructure at megabits per second, provides a paradigm that is impossible for a driver – active or standby – to monitor. Particularly when involving manoeuvres for several vehicles (e.g. for collision avoidance or ramp management).

Casner and his colleagues point out that almost anything a driver does in such situations is likely to degrade the automatically computed solution. They question how we manage *'the crossover point where automation systems are not yet robust and reliable enough to operate without humans standing by to take over but yet are too complex for people to comprehend'.*

With respect to the situation where drivers are unexpectedly asked to reassume control of the car, Casner and his colleagues point out that the driver is going to struggle to get back 'in the loop' to assess the situation and be able to respond in time. They observe that some of these struggles will arise from having to gather the details of the vehicle's situation, while others arise from the complexity of the automation itself – when the details of how the automation works will likely elude the driver's understanding.

By the time we reach Level 4 and 5 vehicles, full automation means exactly what it says on the label. Therefore, there are no issues surrounding a fallback driver. The vehicle will drive, and in the event of an unmanageable issue, will stop safely.

Casner et al. conclude that 'due to the many remaining obstacles, and the rate at which cars are replaced on the roadways worldwide, the transition to fully automated driving for a majority of the public will take decades. The safety challenges of partially automated driving will be significant and, at least as of today, underestimated. .... Today, we have accidents that result when drivers are caught unaware. Tomorrow, we will have accidents that result when drivers are caught even more unaware. We can only echo a plea that is being made to drivers today: Set personal electronic devices aside, resist any temptation to become distracted, and remain focused on the road.'

Casner et al. opine: *'We expect the most serious problems to arise in systems that take the driver out of the loop, yet these are the very systems drivers want, precisely because they free the driver to do something other than drive.'*

This author adds advice to regulators to heed the evidence presented. To ignore the hype pushing to rapid instantiation of Level 3 systems, and err on the side of safety and caution, not the ambition of automotive companies.

To vehicle system designers a concluding suggestion from Casner and colleagues for DAS is 'To help maintain driving skill, wakefulness, or attentiveness, car interfaces might periodically ask the driver to assume manual control.'

NHTSA, the US National Highway Traffic Safety Administration, acknowledging the work of Casner et al., identify the following key issues regarding 'human factors problems' associated with different levels of automated driving. For understanding, you may assume that the NHTSA levels of automation equate to the J3016 levels (Table 8.1).

What is clear from this and other work is that the design of the HMI is critical. Fortunately, automotive manufacturers have a lot of experience here. Such systems will have to clearly inform the driver (visually, haptically and acoustically) with sufficient advance time to be able to resume the driving task.

A study by Kühn (2016) determined that, in the use case of automated driving, an additional five seconds more than typical three- to four-second response time for conventional driving was needed, which means that nine seconds is needed before the driver is fully back in control and aware. Even drivers who were not distracted at the moment when the request of intervention came had delayed reactions compared to users in normal manual driving.

Kühn established that, after driving with a high level of distraction, 90% of the drivers looked at the road again for the first time after three to four seconds, had their hands back on the steering wheel and their feet on the pedals after six to seven seconds and had turned off the automated system after seven to eight seconds.

Table 8.1 Human factors.

Level of automation (NHTSA)	Human factors problems
1. (e.g. adaptive cruise control)	Vigilance: taking drivers out of the active control makes it difficult to get them back in when it is necessary, as previous studies have reported reduced vigilance, increased drowsiness and longer reaction times to unexpected events when relieving drivers of even one aspect of the driving task (Dufour, 2014 as cited in Casner et al. (2016)).
2. (e.g. traffic jam assist, park assist Level 2)	Inattention: as automation becomes more able and reliable, drivers will inevitably do things other than pay attention to driving. Feedback: knowing the state of the automation is of paramount importance and this is not straightforward. On the one hand, users rely on their memory of having pushed a button and habitually ignore system-status displays. On the other hand, automation functions sometimes turn off without any apparent reason and lacking an appropriate feedback.
3. (e.g. traffic jam chauffeur, highway chauffeur, highway pilot)	Rapid onboarding: users have great difficulty re-establishing driving context and this is especially worse when the situation is complex. Skill atrophy: cognitive skills deteriorate when not practiced regularly but hands on skills seem to be resistant to forgetting Casner et al. (2016). However, cognitive skills are needed first to determine what manual operations are required. Complexity: drivers are less trained compared to pilots of an airplane, which creates critical situations where the automation complexity results in unexpected behaviours. When drivers are unexpectedly asked to resume control of the car, they are likely to experience difficulties to get back in the loop, assess the situation and be able to respond in time.

Kühn recommended:

- The driver needs to be notified as early and clearly as possible of the need to resume vehicle control (preceded by an early identification of the need to transfer the vehicle control). The takeover period must last longer than eight seconds.
- The automated system must remain active during the takeover process, until the driver has shown readiness to take over vehicle control.
- A minimum risk manoeuvre has to be put in place if the driver cannot handle the control takeover request.
- Comprehensive but succinct information on the current driving situation needs to be provided in order to facilitate the driver's situational awareness after an automated drive.
- The vehicle readiness to assist after the driver has resumed control needs to be increased to avoid inappropriate reactions of the driver.
- To show the urgency of a given situation, a cascade of different types of warnings could be issued to the driver.

- Instructions about capabilities and limitations of automated systems could be specifically given to drivers for better user reactions in the event of a control takeover request and to avoid wrong or reduced system use.

That said, the overriding concern remains. In the event of a road incident such as an impending crash situation, empirical and test environment data shows that there is commonly only one to four seconds notice of an incident/crash, sometimes less, especially at highway speeds, before the event/impact. Yet, as we have seen above, it takes eight to nine or more seconds for the driver to reconnect with the driving task, if the AV system requests the driver to take over. As they say popularly in sci-fi stories 'This does not compute' and as will be commented in social media post event 'how well did that work out for you?'

8.21 Range of Services Supported

> *The operator of the infrastructure C-ITS station will then have to identify what ITS services it will support and how. No detailed system analysis of any of these services has been circulated, and no standard service definition issued. These steps need to happen very quickly. And these services need to be defined clearly, including data tree definition and unambiguous data definition, probably in ASN.1.*

Which comes first, the chicken or the egg?

Equipping vehicles with wireless connectivity is only an attractive proposition if that connectivity enables new services to the vehicle owner/keeper/driver/passengers versus what is the point of providing services before vehicles are connected?

The decision by Volkswagen (Business Insider 2019/10) to adopt the EC and ITS sector preferred route, and the decision of others, such as Volvo, to follow, at least for Level 1 and 2 automation, should hopefully break the logjam and now enable rapid implementation of 'connected' vehicles. But what services will be supported? From earlier chapters the reader will be aware the EU, through the European Commission inspired C-ITS platform, identified 14 'Day 1 services' and 7 'Day 1.5 services'. The C-ITS Platform report (together with the subsequent C-ROADS implementation initiative), which has largely formed EC policy for connected roads, states the intention 'These applications should be the first to be implemented in the EU by 2020.'

List of Day 1 Services

- Hazardous location notifications:
- Slow or stationary vehicle(s) and traffic ahead warning
- Road works warning
- Weather conditions
- Emergency brake light
- Emergency vehicle approaching
- Other hazardous notifications
- Signage applications:

- o In-vehicle signage
- o In-vehicle speed limits
- Signal violation/intersection safety
- Traffic signal priority request by designated vehicles
- Green light optimal speed advisory (GLOSA)
- Probe vehicle data
- Shockwave damping (falls under ETSI category 'local hazard warning')

List of Day 1.5 Services

- Information on fuelling and charging stations for alternative fuel vehicles
- Vulnerable road user protection
- On-street parking management & information
- Off-street parking information
- Park and ride information
- Connected & cooperative navigation into and out of the city (first and last mile, parking, route advice, coordinated traffic lights)
- Traffic information and smart routing

8.21.1 Services that Can Be Instantiated Without the Support of the Local Infrastructure

Without going too deeply into the system design and architecture options, of these services, the following can be provided (at least to a minimum service workable provision level) without the support of the local infrastructure:

(Day 1)

- Hazardous location notifications (within 500 m) by V2V communications
- Slow or stationary vehicle(s) and traffic ahead warning (within 500 m) by V2V communications
- Road works warning (within 500 m) by V2V communications
- Weather conditions via e.g. 4G weather service (Accuweather-type enquiry from vehicle) or weather hazard warnings (within 500 m) by V2V communications (forwarding ice/skid alerts from traction control events, or wiper activation events from one vehicle broadcast to nearby vehicles)
- Emergency brake light (within 500 m) by V2V communications
- Emergency vehicle approaching (within 500 m) by V2V communications
- Signage applications: from in-vehicle sensors reading road signs
  - o In-vehicle signage
  - o In-vehicle speed limits

(Day 1.5)

- Vulnerable road user protection *from in-vehicle sensors and (within 500 m) by V2V communications*

8.21.2 Services that Can Only Be Provided Using Data/Information from the Local Infrastructure

The following list of services can only be provided using data/information from the local infrastructure:

(Day 1)

- Road works warning
- Other hazardous notifications
- Signal violation/intersection Safety
- Traffic signal priority request by designated vehicles
- Green light optimal speed advisory (GLOSA)
- Probe vehicle data
- Shockwave damping (falls under ETSI category 'local hazard warning')

(Day 1.5)

- Information on fuelling and charging stations for alternative fuel vehicles
- On-street parking management and information
- Off-street parking information
- Park and ride information
- Connected and cooperative navigation into and out of the city (first and last mile, parking, route advice, coordinated traffic lights)
- Traffic information and smart routing

8.21.3 Services that Can Be Enhanced/Improved/Extended by Using Data/Information from the Local Infrastructure

The following list of services can be enhanced/improved/extended only by using data/information from the local infrastructure:

(Day 1)

- Hazardous location notifications *(Providing localised local traffic management centre [TMC]/Transport optimisation service [TOS] data to vehicles via roadside C-ITS station)*
- Slow or stationary vehicle(s) and traffic ahead warning *(Providing localised local TMC/TOS data to vehicles via roadside C-ITS station)*
- Road works warning *(Providing localised data to vehicles via roadside C-ITS station)*
- Weather conditions *(Broadcasting local weather forecasts to vehicles via roadside C-ITS station; forwarding collected ice/skid/rain warnings to ensuing vehicles via roadside C-ITS station)*
- Emergency vehicle approaching *(Managing the journey of the emergency vehicle/ monitoring its progress, and providing timely notification warnings to vehicles via roadside C-ITS station)*
- Other hazardous notifications *(Providing localised collected data + TMC/TOS data to vehicles via roadside C-ITS station)*
- Signage applications:

- o In-vehicle signage *(Providing localised electronic traffic regulation information [METR] to vehicles* via *roadside C-ITS station)*
- o In-vehicle speed limits *(Providing localised electronic traffic regulation information [METR] to vehicles* via *roadside C-ITS station)*
- Signal violation/intersection safety *(Infrastructure equipped with C-ITS stations to communicate directly with equipped vehicles)*
- Traffic signal priority request by designated vehicles *(Infrastructure equipped with C-ITS stations to communicate directly with equipped vehicles)*
- Green light optimal speed advisory (GLOSA) *(Infrastructure equipped with C-ITS stations to communicate directly with equipped vehicles using TMC/TOS data)*
- Probe vehicle data *(Infrastructure equipped with C-ITS stations to collect data from C-ITS equipped vehicles, and pass probe data results to equipped vehicles upon enquiry or as localised broadcast information* via *C-ITS stations)*
- Shockwave damping (falls under ETSI category 'local hazard warning') *(Providing localised collected data + TMC/TOS data to vehicles* via *roadside C-ITS station)*

(Day 1.5)

- Information on fuelling and charging stations for alternative fuel vehicles *(Providing localised fuelling & charging stations information [including dynamic usage data] to vehicles* via *roadside C-ITS station)*
- Vulnerable road user protection *(Providing localised collected data + TMC/TOS data to vehicles* via *roadside C-ITS station)*
- On-street parking management and information *(Providing localised static and dynamic/TOS data to vehicles* via *roadside C-ITS station)*
- Off-street parking information *(Providing localised static and dynamic/TOS data to vehicles* via *roadside C-ITS station)*
- Park and ride information *(Providing localised static and dynamic/TOS data to vehicles* via *roadside C-ITS station)*
- Connected and cooperative navigation into and out of the city (first and last mile, parking, route advice, coordinated traffic lights) *(Providing localised communication with TOS and/or + dissemination of information from TMC to vehicles* via *roadside C-ITS station)*
- Traffic information and smart routing *(Providing localised communication with TOS and/or + dissemination of information from TMC to vehicles* via *roadside C-ITS station)*

At least within the EU, as it is with the European Commission policy endorsed by Member States, road operators/local roads authorities should at the least be prioritising the development and support of the services listed in Section 8.21.2.

If road operators, and particularly local roads authorities, are seriously committed to use cooperative ITS to assist their struggle to reduce road deaths and injuries, they should urgently be prioritising the development and support of the services listed in Section 8.21.3 above.

If the national government are seriously committed to use cooperative ITS to assist their struggle to reduce road deaths and injuries, they should urgently be prioritising the development and support of the services listed in Section 8.21.3 above, and considering what steps they can take to encourage and support road operators/local authorities to provide such service support.

Venture capitalists/investors need to examine business cases to see how they can monetise this opportunity.

The European Commission and US DoT/FHWA should consider what assistance/harmonisation they can provide, as a matter of urgency. It does not make sense for every local authority to determine these services differently. It would make sense for the European Commission and US DoT/FHWA to at least develop and provide common basic system specifications, and even to provide software specifications and fund the development of associated standards.

A first attempt to determine an architecture for these services has been made in the 'Harmonization Task Group' (HTG) (a collaboration of US DOT, the EU JRC, EC DG, EC DG GROW, CNECT and the TCA of Australia), by the development of the HARTS ITS architecture. The extracts below provide that (internationalised) architecture for the EC's list of Day 1 and Day1.5 services, which could provide a start point for this work.

8.21.4   The HARTS Architecture with Reference to C-ITS Platform Day/Day 1.5 Services

In November 2009 the United States Department of Transportation (USDOT) and Directorate General for Information Society and Media (DG INFSO) signed a Joint Declaration of Intent on Research Cooperation. The goal of the declaration is to:

> *Support, wherever possible, global open standards in order to ensure interoperability of cooperative systems worldwide and to preclude the development and adoption of redundant standards.*

This harmonisation effort is intended to contribute to deployment of interoperable cooperative vehicle and infrastructure systems. 'Standards harmonization' is a process through which various stakeholders – vehicle and equipment manufacturers, technical standards development organisations (SDOs), and governments – work together to achieve the optimum level of harmonisation needed for efficient deployment of cooperative ITS.

The United States (US) and European Union (EU) participate in international standards harmonisation activities focusing on technical standards 'around the vehicle platform,' that is, standards needed to provide connectivity between vehicles and between vehicles and infrastructure. Harmonisation research for ITS technical standards complements, and will be a key enabler for, the efficient deployment of V2V and V2I technologies.

Development and adoption of coordinated harmonised international technical standards contribute to the following benefits:

• Improved interoperability and interchangeability of intelligent transportation systems (ITS) across operational boundaries;

- Reduced development and deployment costs for manufacturers;
- Greater accessibility to international markets for manufacturers of connectivity equipment;
- Increased competition and innovation amongst manufacturers that can help lower costs and expand service for consumers;
- The potential for a more rapid deployment of ITS systems;
- Leveraging of international expertise and reducing redundant efforts.

The 'harmonization task group' (HTG) has produced valuable work over the past decade, working to try to find consensus, and more importantly, trying to prevent divergence, between the continents, as C-ITS evolves and is deployed. The HTG is now a collaboration of USDOT, the EU JRC, EC DG, EC DG MOVE, CNECT and the TCA of Australia. (Japan also participates, but is ploughing a different furrow in practice.)

One of HTG's major achievements has been to bring together the ITS architectures from the United States, Europe, Australasia and with the help of experts around the world, and fuse them the HARTS ITS Architecture, which can be viewed at http://htg8 .org/index.html (n.d., 2019)

The following extracts show the HARTS architecture view, as of November 2019, or from ARC-IT version 8.3 https://local.iteris.com/arc-it in respect of the C-ITS platform Day 1/Day 1.5 services – covering much of the J 3016, DAS Level 1–3 services (n.d., 1945)

8.21.4.1   Hazardous Location Notifications
See Figures 8.9–8.11.

8.21.4.2   Slow or Stationary Vehicle(s) and Traffic Ahead Warning
See Figures 8.12 and 8.13.

8.21.4.3   Road Works Warning
See Figures 8.14 and 8.15.

8.21.4.4   Weather Conditions
See Figure 8.16.

8.21.4.5   Emergency Brake Light
See Figure 8.17.

8.21.4.6   Emergency Vehicle Approaching
See Figure 8.18.

8.21.4.7   Other Hazardous Notifications and Shockwave Damping (Falls Under ETSI Category 'Local Hazard Warning')
See Figures 8.19 and 8.20.

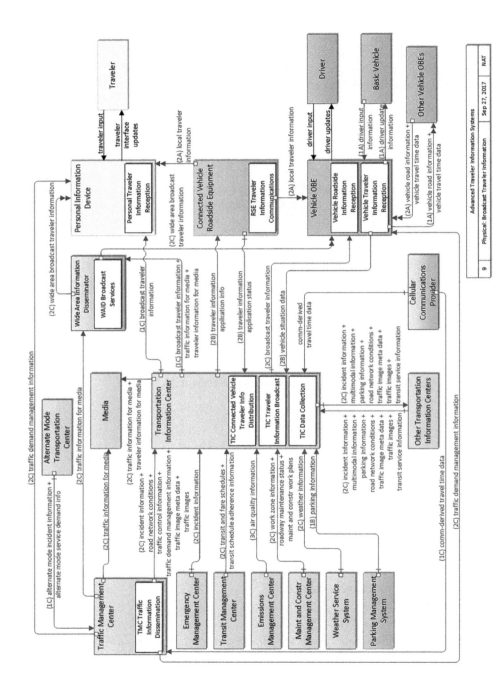

Figure 8.9 Advanced traveller information systems. Source: Courtesy of Harmonized Architecture Reference for Technical Standards.

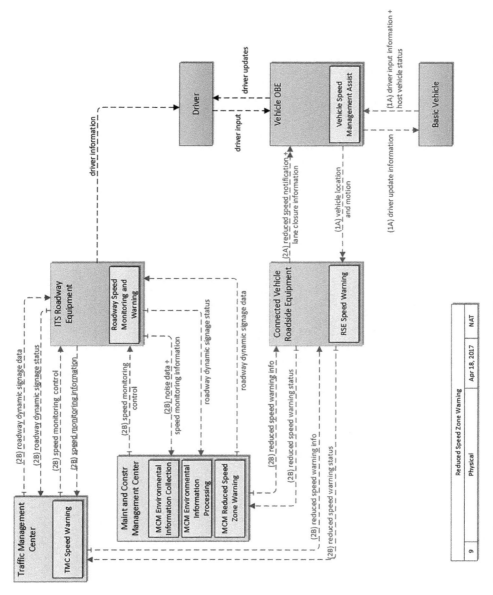

Figure 8.10 Reduced speed zone warning/lane closure. Source: Courtesy of Harmonized Architecture Reference for Technical Standards.

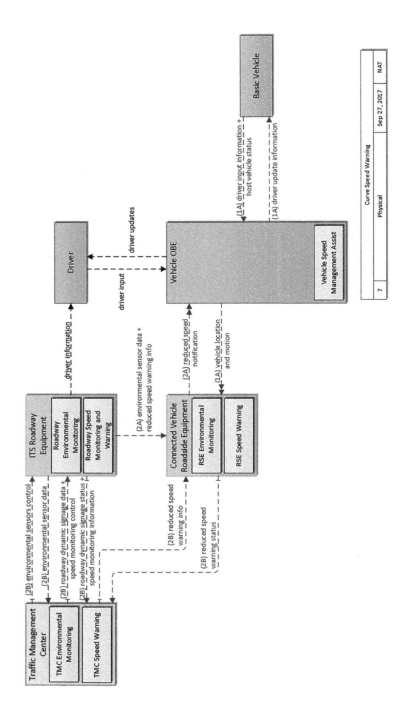

Figure 8.11 Curve speed warning. Source: Courtesy of Harmonized Architecture Reference for Technical Standards.

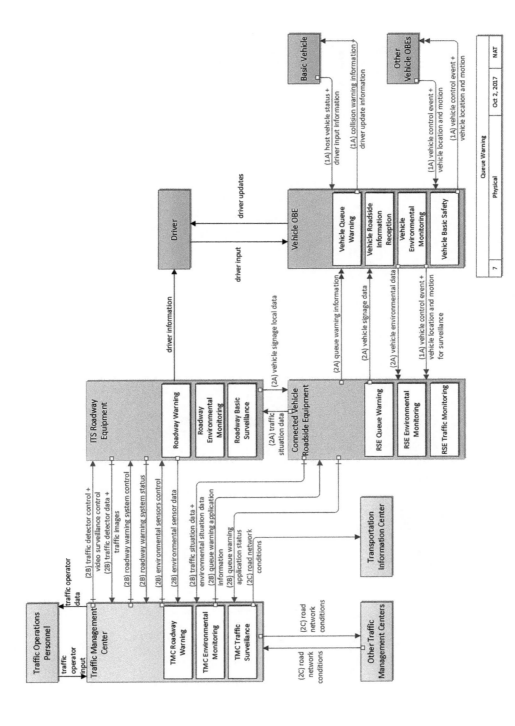

Figure 8.12 Queue warning. Source: Courtesy of Harmonized Architecture Reference for Technical Standards.

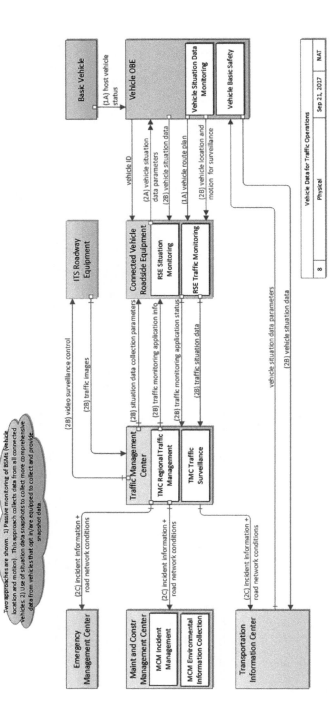

Figure 8.13 Vehicle data for traffic operations. Source: Courtesy of Harmonized Architecture Reference for Technical Standards.

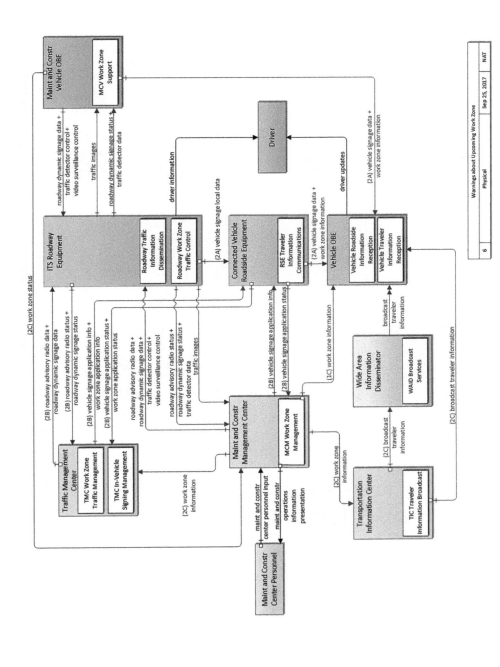

Figure 8.14 Warnings about upcoming work zone. Source: Courtesy of Harmonized Architecture Reference for Technical Standards.

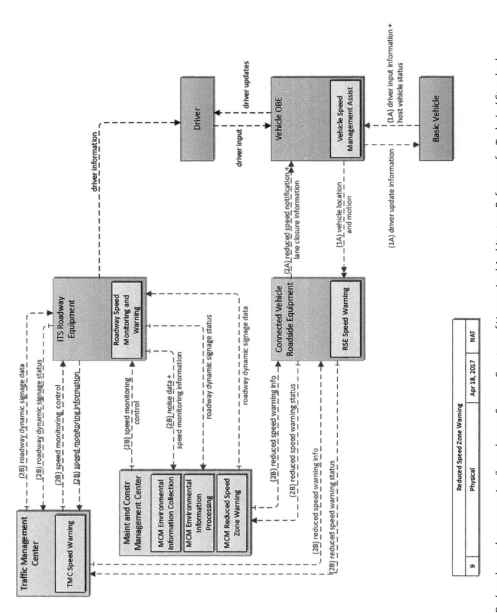

Figure 8.15 Reduced speed zone warning/lane closure. Source: Courtesy of Harmonized Architecture Reference for Technical Standards.

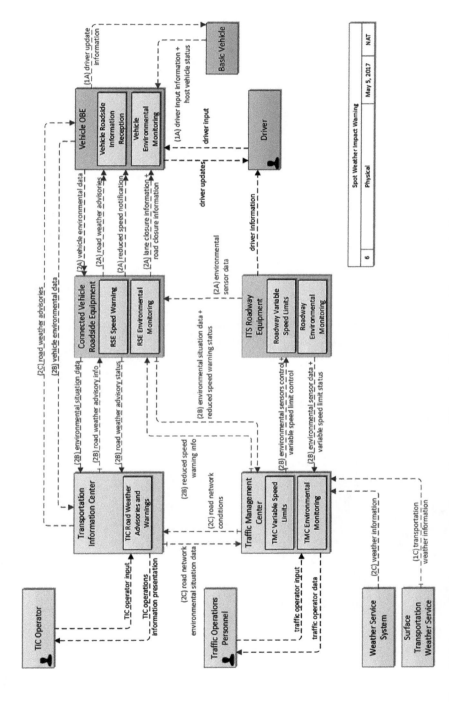

Figure 8.16 Spot weather impact warning. Source: Courtesy of Harmonized Architecture Reference for Technical Standards.

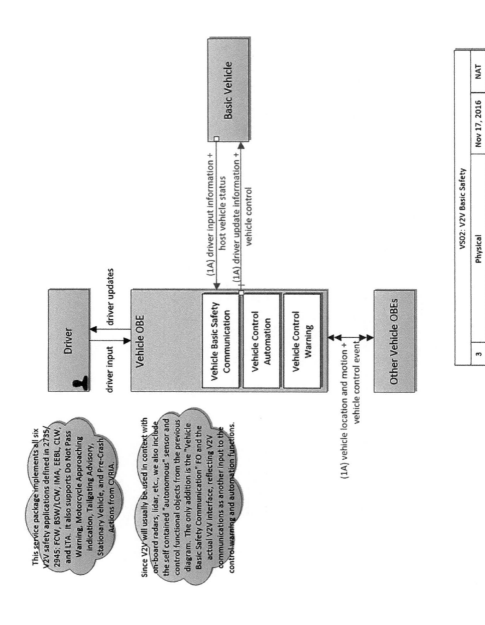

Figure 8.17 Emergency brake light. Source: Courtesy of Iteris. ARC-ITS Architecture.

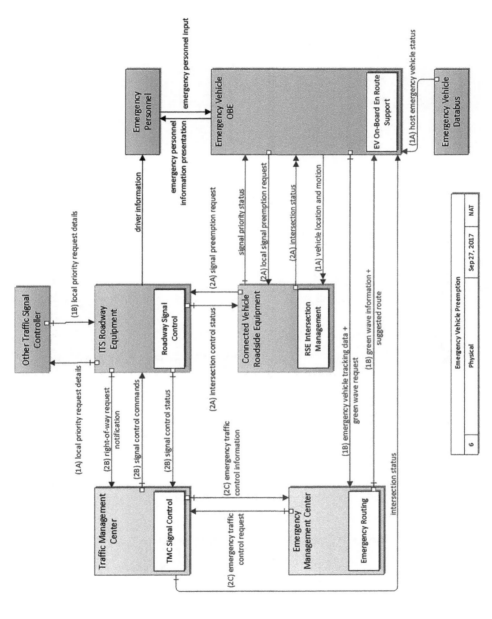

Figure 8.18 Emergency vehicle preemption. Source: Courtesy of Harmonized Architecture Reference for Technical Standards.

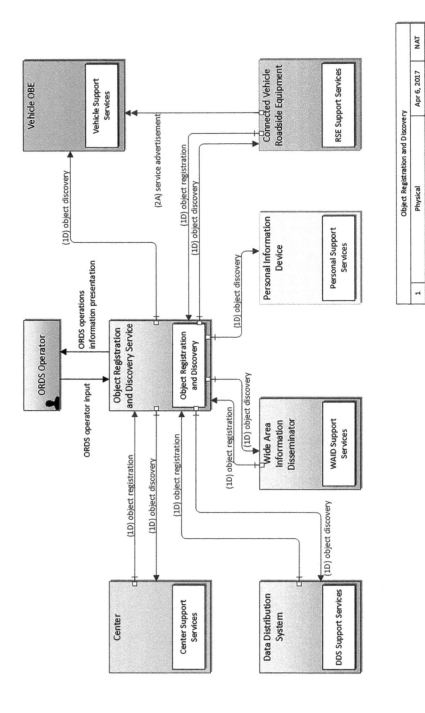

Figure 8.19 Object registration and discovery. Source: Courtesy of Harmonized Architecture Reference for Technical Standards.

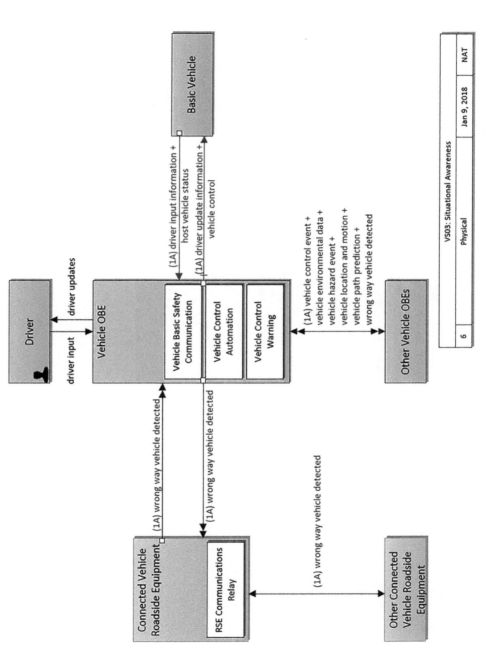

Figure 8.20 Pedestrian in signalised crosswalk warning. Source: Courtesy of Harmonized Architecture Reference for Technical Standards.

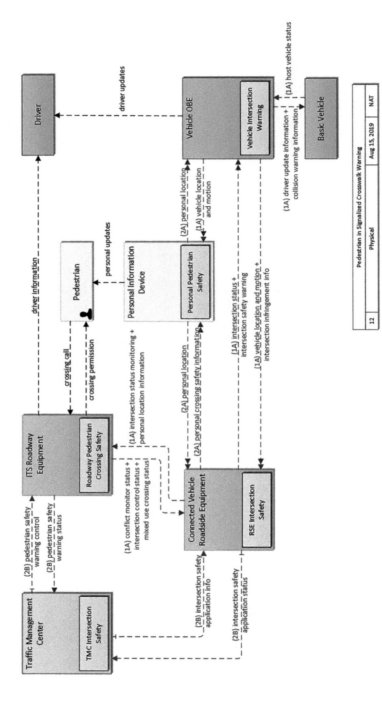

VS03 Situational Awareness (Source: Courtesy of Iteris. ARC-ITS Architecture)

8.21.4.8   Signage Applications: In-Vehicle Signage
See Figure 8.21.

8.21.4.9   Signage Applications: In-Vehicle Speed Limits
See In-Vehicle Signage or
   Advanced Traveller Information Systems

8.21.4.10   Signal Violation/Intersection Safety
See Figure 8.22.

8.21.4.11   Traffic Signal Priority Request by Designated Vehicles
See Figures 8.22 and 8.24.

8.21.4.12   Green Light Optimal Speed Advisory (GLOSA)
See Figure 8.25.

8.21.4.13   Probe Vehicle Data
See Figure 8.26.

8.21.4.14   Information on Fuelling and Charging Stations for Alternative Fuel Vehicles
See Figure 8.27.

8.21.4.15   Vulnerable Road User Protection
See Figures 8.28 and 8.29.

8.21.4.16   On-Street Parking Management and Information
See Figure 8.30.

8.21.4.17   Off-Street Parking Information
See Figure 8.31.

8.21.4.18   Park and Ride Information
See Figure 8.32.

8.21.4.19   Connected and Cooperative Navigation Into and Out of the City
(First and last mile, parking, route advice, coordinated traffic lights)
   See Figures 8.33 and 8.34.

8.21.4.20   Traffic Information and Smart Routing
See Figures 8.35–8.41.

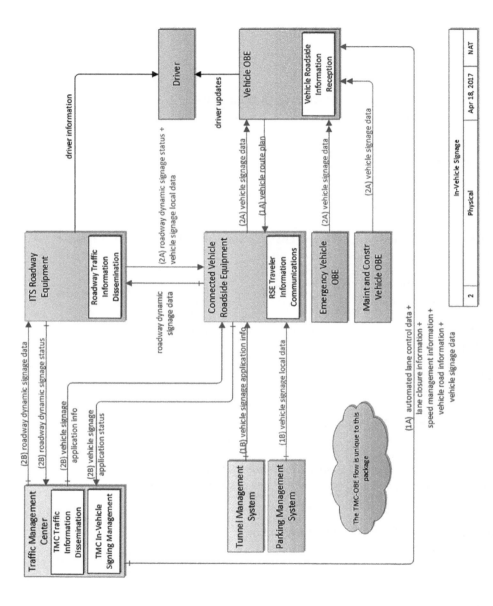

Figure 8.21 In-vehicle signage. Source: Courtesy of Harmonized Architecture Reference for Technical Standards.

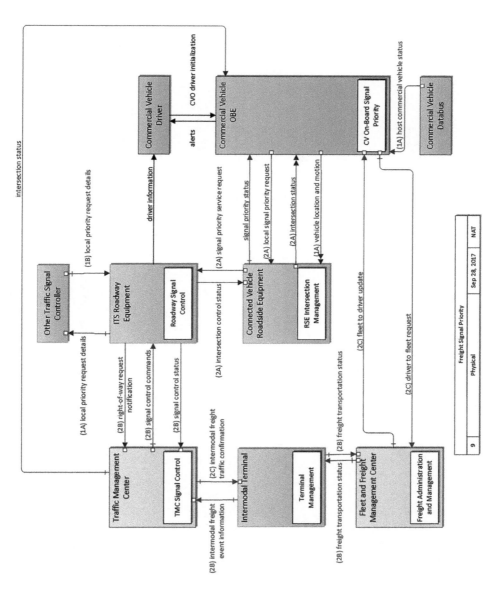

Figure 8.22 Stop sign violation warning. Source: Courtesy of Harmonized Architecture Reference for Technical Standards.

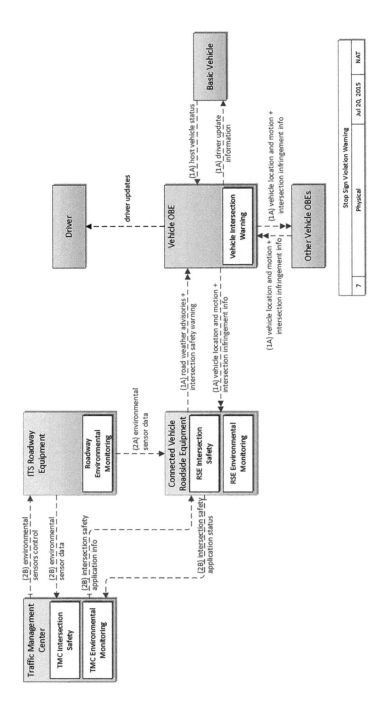

Figure 8.23 Freight signal priority. Source: Courtesy of Harmonized Architecture Reference for Technical Standards.

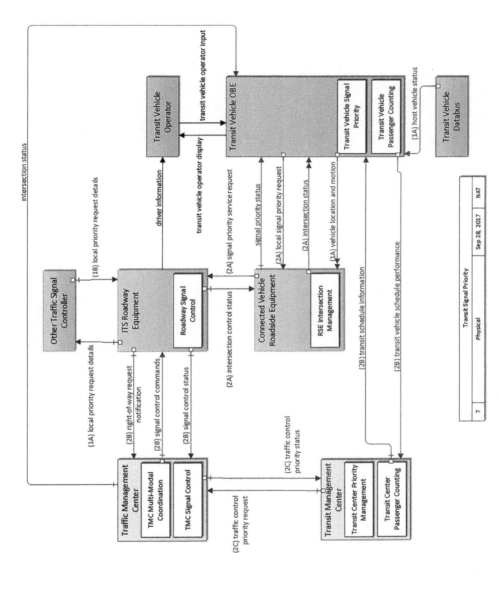

Figure 8.24 Transit signal priority. Source: Courtesy of Harmonized Architecture Reference for Technical Standards.

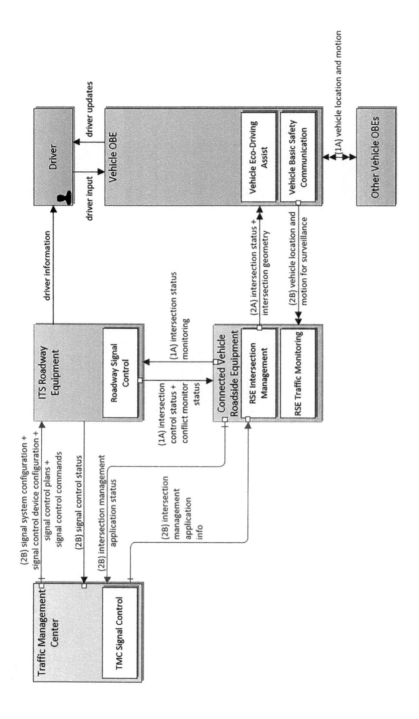

Figure 8.25 Eco-approach and departure at signalised intersections. Source: Courtesy of Iteris. ARC-ITS Architecture.

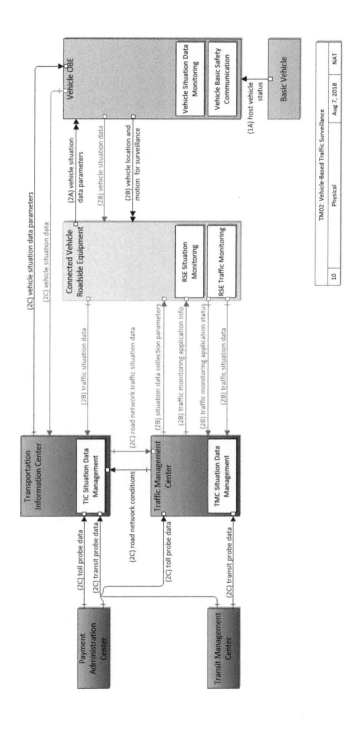

Figure 8.26 TM02 vehicle-based traffic surveillance. Source: Courtesy of Iteris. ARC-ITS Architecture.

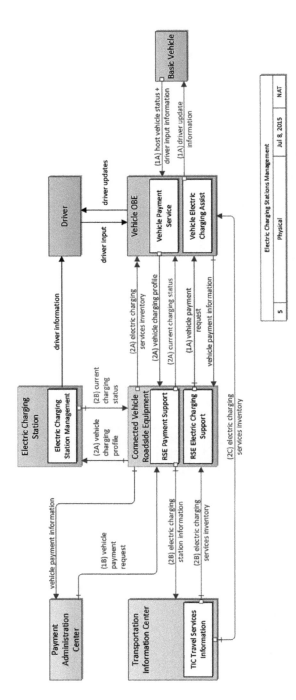

Figure 8.27 Electric charging stations management. Source: Courtesy of Harmonized Architecture Reference for Technical Standards.

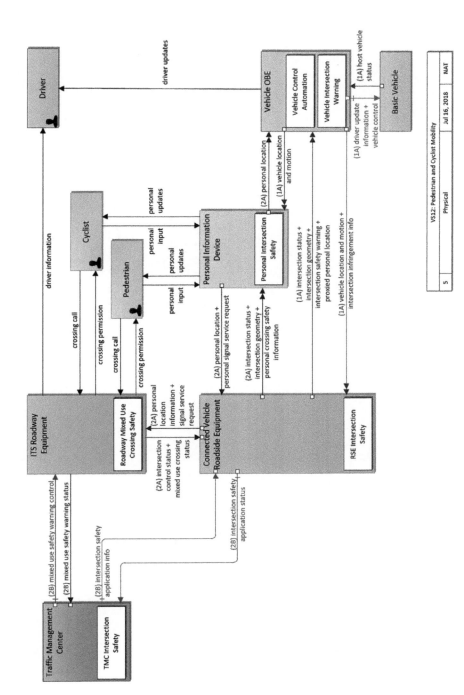

Figure 8.28 VS12 pedestrian and cyclist safety. Source: Courtesy of Iteris. ARC-ITS Architecture.

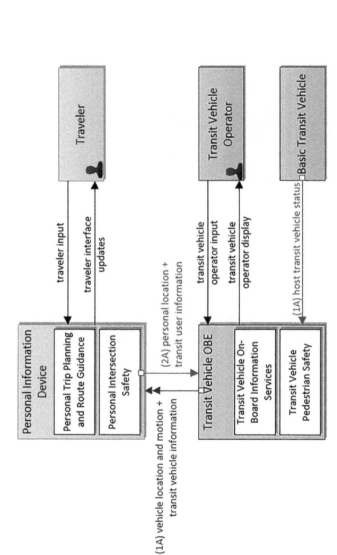

PT11: Transit Pedestrian Indication		
10	Physical	Feb 8, 2017	NAT

Figure 8.29 PT11 transit pedestrian indication. Source: Courtesy of Iteris. ARC-ITS Architecture.

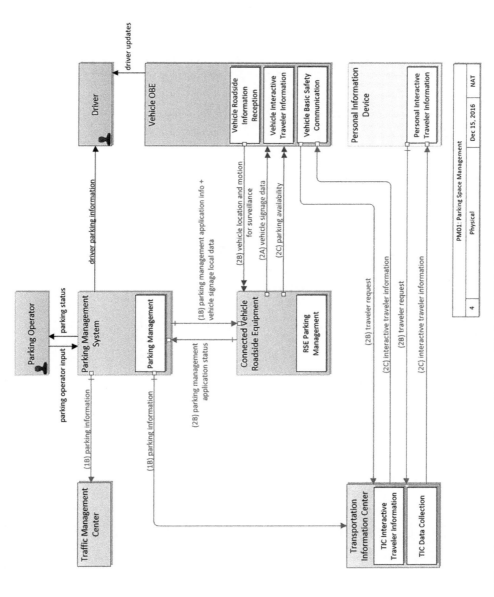

Figure 8.30 PM01 parking space management. Source: Courtesy of Iteris. ARC-ITS Architecture.

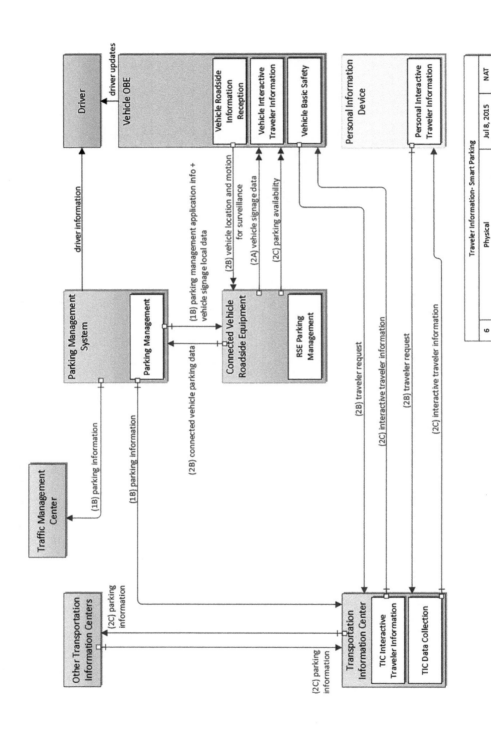

Figure 8.31 Traveller information – smart parking. Source: Courtesy of Harmonized Architecture Reference for Technical Standards.

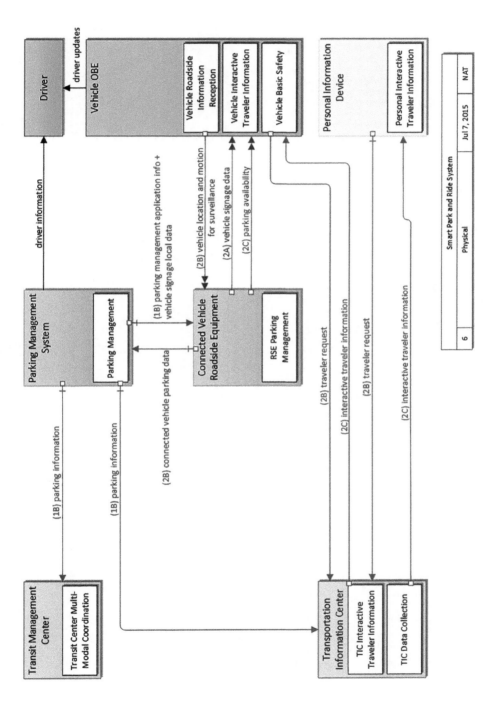

Figure 8.32 Smart park and ride system. Source: Courtesy of Harmonized Architecture Reference for Technical Standard.

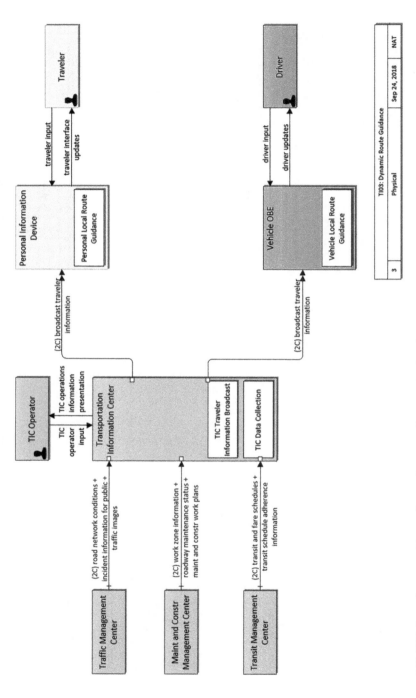

Figure 8.33 TI03 dynamic route guidance. Source: Courtesy of Iteris. ARC-ITS Architecture.

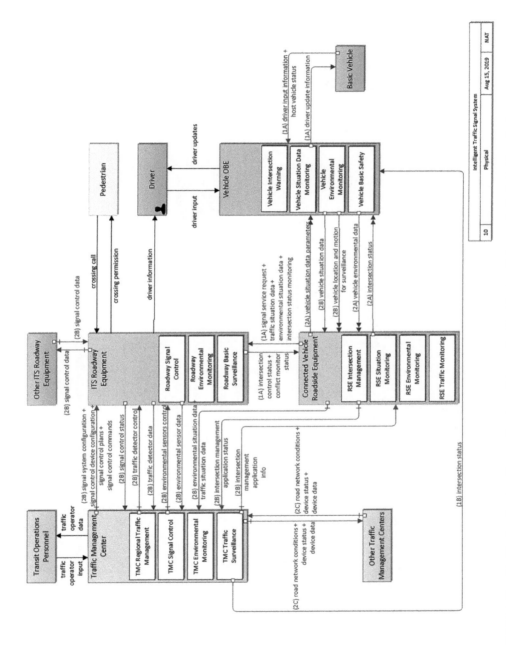

Figure 8.34 Intelligent traffic signal system. Source: Courtesy of Harmonized Architecture Reference for Technical Standards.

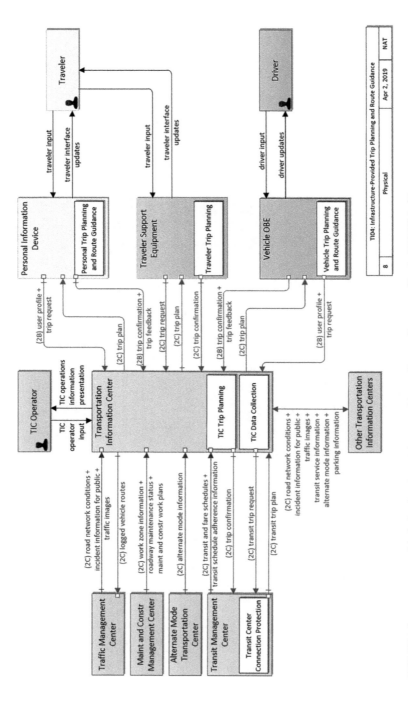

Figure 8.35 I04 infrastructure-provided trip planning and route guidance. Source: Courtesy of Iteris. ARC-ITS Architecture.

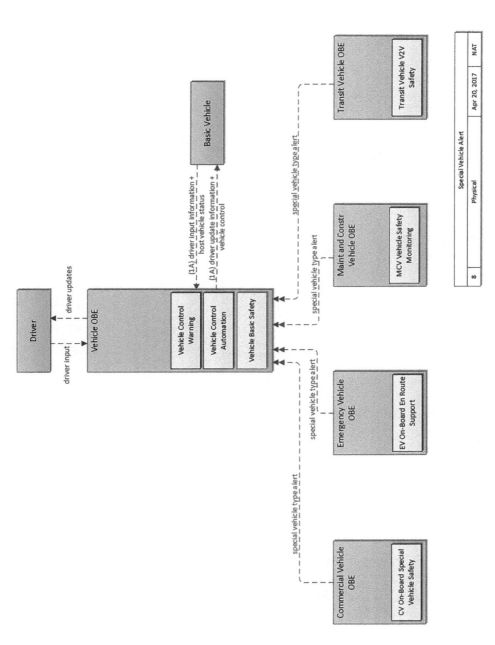

Figure 8.36 Special vehicle alert. Source: Courtesy of Harmonized Architecture Reference for Technical Standards.

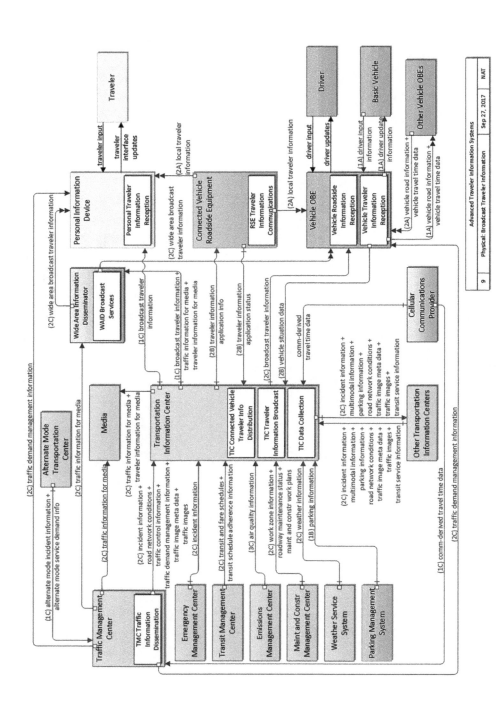

Figure 8.37 Advanced traveller information systems. Source: Courtesy of Harmonized Architecture Reference for Technical Standards.

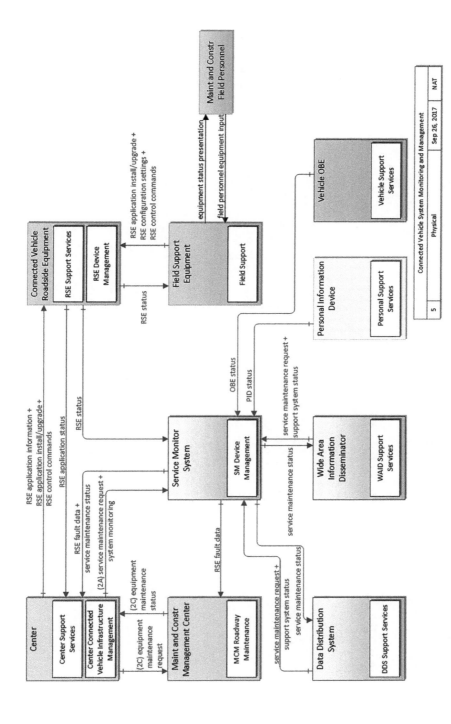

Figure 8.38 Connected vehicle system monitoring and management. Source: Courtesy of Harmonized Architecture Reference for Technical Standards.

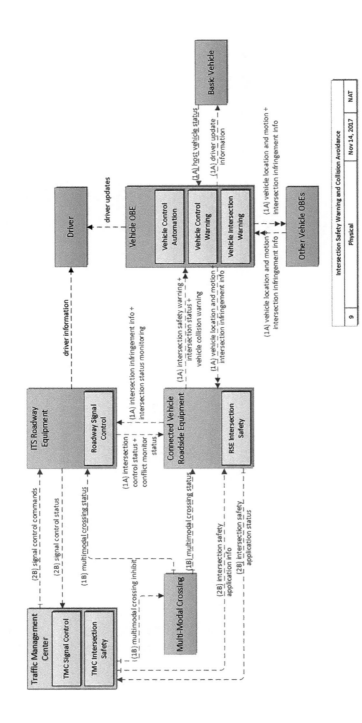

Figure 8.39 Intersection safety warning and collision avoidance. Source: Courtesy of Harmonized Architecture Reference for Technical Standards.

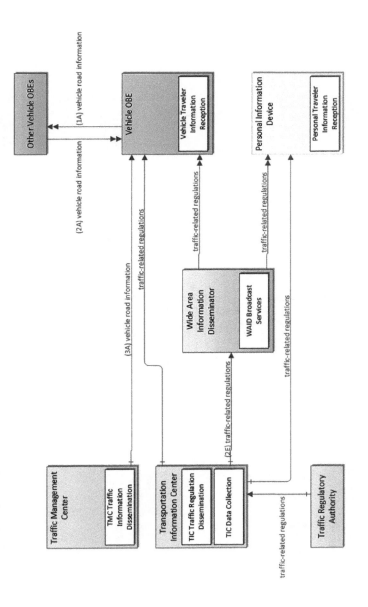

Figure 8.40 Electronic regulations. Source: Courtesy of Harmonized Architecture Reference for Technical Standards.

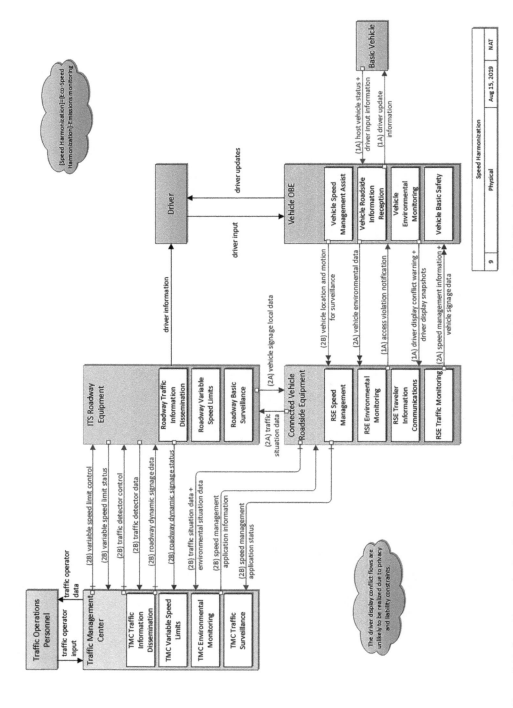

Figure 8.41 Speed harmonisation. Source: Courtesy of Harmonized Architecture Reference for Technical Standards.

8.22 Young Drivers and Experience

Young drivers who have access to automated driving may build up less driving experience. This is an area that needs more research. This also poses questions for driver training: how will training teach people to drive safely and make the most of automated driving techniques, and how will drivers be taught to safely make the switch between fully autonomous and automated driving?

This is an interesting area, because it is an indicator that the advent of AVs will not just replace some or all manually driven vehicles, but will introduce societal changes about the way we travel, and the way we perceive travelling, indeed, the very point of putting the two subject areas of AV and MaaS together in this work is because the advent of AVs fundamentally changes the way we travel and how the MaaS paradigm is instantiated.

The points of concern expressed above require further research, especially, as it becomes clearer as we work through the research and practical finding of this book, that the period of DAS systems, before Level 4 and 5 will be allowed and instantiated on our roads, is likely to be in excess of a decade, and well beyond that before automated vehicles predominate the roads.

The consequences are that for the next quarter century at least, young persons will be exposed to the manual driving paradigm, much as they are today. They will sit the driving test and physically gain experience, first as passengers, secondly through driving lessons. The fear expressed in the opening statement is valid, but in respect of AVs it is valid in one or two decades' time.

And to some extent the concern is probably groundless. Young people today, all over the world, have access to gaming stations, or have games on smartphones (at least those in car-driving societies). One of the most popular categories of video games is car racing of some description. A growing number of these games are 'virtual' reality experiences. Indeed, one of the problems today is that, thanks to video games, young teenagers are capable to get into a car and drive it without the help of Mum or Dad, or lessons from the driving school. And then drive it aggressively as they do in their video games, often with tragic results.

So, the probability is that, long after young persons *need* to be able to drive (because the roads have become automated), courtesy of their virtual reality gaming devices, they will continue to do so for fun. Indeed, once AV becomes predominant, the idea of 'taking control of the driving experience' in VR games is likely to become even more attractive.

But the opening contention, in combination with Maas, does beg a valid question in the next decade. If *MaaS* becomes popular, increasingly fewer young people will want to learn to drive. While the AV turning up at the front doorstep may be quite a few years away, eScooter share services, shared bicycle services, ride hailing, all present young people with nondriving options for the last mile. Combined with the weight of student debt, and difficulty affording rent or saving for house purchase, the availability of these MaaS options will incline many away from car ownership and its associated costs (which will only rise), especially in conurbations. Then, in this decade we do have the problem of DAS vehicles, and potential users without the skills to drive them, which is a business opportunity perhaps for those providing virtual reality video games.

How will training teach people to drive safely and make the most of automated driving techniques, and how will drivers be taught to safely make the switch between fully autonomous and automated driving? Another business opportunity for those providing virtual reality video games.

8.23   Liability

Liability in the event of a collision with an automated vehicle needs to be more clearly defined and regulated.

The liability in the event that things go wrong remains very unclear, and until these issues are clarified, vehicle manufacturers may be loth to provide AV as an option, and governments could find themselves at the wrong end of expensive legal actions. Venture capitalists and investors are also likely, and are advised to, take these risk issues carefully into account before committing.

Over time there will need to be a fundamental change from a responsible named driver having responsibility to a situation where the manufacturer will be held accountable. Indeed, some of the largest manufacturers (Volvo, Tesla, and others) have already indicated that they are so confident in their technology that they will accept the liability.

This will be bad news for independent garage owners, with manufacturers set to insist all services and repairs must be provided by their own garages. And there is a wider issue. How does this fit with EU's single open market? It will lock vehicle users into monopolistic service provision. Further, if your insurer is also the technology provider, they are an interested party. If the issue in a claim is 'Is the driver to blame or is the AV system to blame?', it will be difficult for them to be objective. And during the transition, period, especially the DAS levels, of automation, the question of responsibility will be difficult to answer.

The situation is unclear in all countries. I will use UK as an example. As stated above UK has passed 'The Automated and Electric Vehicles (AEV) Act, 2018'.

The automated vehicle (part 1), of the Act applies to self-driving vehicles that '*may lawfully be put in self-drive mode on roads or other public places in Great Britain*'.

The key provision within part 1 of the Act is section 2, which states that an insurer will be directly liable for an accident caused by an AV where it is:

'driving itself on a road or other public place in Great Britain';
'insured'; and
an 'insured person or any other person suffers damage as a result of the accident'.

This section (S2) is intended to give innocent victims a direct right of action against the vehicle's insurer. The intention is for victims of accidents involving AVs to obtain compensation quickly and easily, without prolonging the process with complicated product liability claims against the AV technology manufacturers, or dealing with liability disputes between insurer and manufacturer.

The insurer then remains free to pursue the manufacturer for any reimbursement or contribution if it can establish that they are liable for the accident in question.

Unfortunately, taken at face value, this wording means the Act does not cover those injured in AV motoring accidents abroad. The duty for compulsory insurance within the UK under Part VI of the Road Traffic Act 1988, extends only to motor vehicles on a road or other public place in Great Britain.

The Act determines that 'driving itself' means 'operating in a mode which is not being controlled and does not need to be monitored by an individual' and clarifies that there is a direct right of action against the insurer where 'an accident is caused by an automated vehicle driving itself [on a road or other public place in Great Britain]'.

But the Act makes no mention of the varying levels of automation at which an AV may be driven, and the different circumstances that may be incurred. In discussions in parliament when the Act was being processed, Baroness Sugg, under Secretary of State for Transport, confirmed that the government would be seeking to set safety standards by way of adopting a UNECE Regulation. (But as we have seen above, UNECE is not yet anywhere near considering regulations for Level 4 and 5 ADS systems.)

She confirmed to the House that 'draft legislation regarding levels of automation will be put before Parliament once there is a better view of the landscape of the new technology and vehicle standards. This would form part of a wider government regulatory programme to ensure motorists and businesses benefit from AVs.'

In respect of 'The Automated and Electric Vehicles (AEV) Act, 2018', this then raises the question, as to which level would be deemed lawful or safe for the purposes of AV mode and, indeed, for the Act to apply??

It is a significant omission that the Act excludes the current and near future semi-autonomous vehicles in which the driver is expected to be monitoring the vehicle whilst in automated mode – which, as the first types of automated driving that will be on the roads, are the highest priority use cases that have need of clarification.

This omission from the scope of the Act suggests that accidents involving these vehicles may have to be resolved by the courts, in cases that may well require complex technical evidence and involve the manufacturer. These questions remain grey areas of the law.

The Act also specifies exclusions on liability to an insured person in defined circumstances, such as when the accident occurs:

- as a direct result of software alterations made by the insured person, or with the insured person's knowledge, that are prohibited under the policy; and
- a failure to install safety-critical software updates that the insured person knows, or reasonably ought to have known, are safety critical.

This then raises a question concerning how much the driver is expected to know about the technology in the vehicle?

Who determines what is 'safety critical'? And who is responsible for informing the driver that there is a safety critical update due? Does this responsibility lie with the vehicle manufacturer, servicing garage or the insurer? And what will be the limitation of insurance cover if the vehicle owner/keeper/driver do not comply? And is the responsibility that of the vehicle keeper, vehicle owner, or the driver?

Is it enough for the keeper or driver to take their car regularly to a reputable garage for servicing? Does this not unreasonably tie them to obtaining servicing from the car manufacturer's agent garage? And, if so, is this not a restraint of trade, illegal under EU single market regulations?

And the consequence is that insurers will likely want to make investigations into the potential exclusions before agreeing to compensate, which may delay the injured party gaining access to compensation at the earliest stage.

The Act (S3(2)) states that the insurer is not liable to the person 'in charge of the vehicle' where the 'accident that it caused was wholly due to the person's negligence in allowing the vehicle to begin driving itself when it was not appropriate to do so'.

But when is it 'appropriate' to allow the vehicle to 'drive itself'? – And who determines this. The Act does not define, but implies, that the driver of the vehicle should be able to judge when it is 'appropriate'. But without the driver being an expert in the strengths, weaknesses and limits of the particular AV system (and they all seem to be different), how can the layman make such judgements?

And the Act is particular in its reference to 'in allowing the vehicle to begin driving itself'. That part – to *begin* driving itself – is clear. The vehicle user should not put the vehicle into AV mode if the conditions are unsuitable. That is logical. But the contentious issues of liability/responsibility are far more likely to be where the vehicle is already operating in DAS mode, and the driver, seeing a risk, does not take-back control, or the AV system demands a fall-back to manual driving, but this does not happen, or does not happen in time to prevent an incident.

There will need to be further guidance on this to ensure injured drivers are afforded sufficient protection. And what are the provisions and realities of taking back control? (see Section 8.20).

And the Act is unclear about who is 'in charge of the vehicle', when a vehicle is driving in automated mode. Is it a 'driver' in the vehicle- in which case there must always be a driver in the vehicle qualified and competent and alert enough to make such decision? Is it the operator of the AV system? (who may or may not be in the vehicle). Is it the designer of the AV system? All of this is left unclear.

The Act confirms that incidents involving AVs will be subject to the usual rules of contributory negligence as defined in the Law Reform (Contributory Negligence) Act 1945 (2019), where the accident or damage was to any extent caused by the injured party.

Therefore, what is the responsibility of a pedestrian, or driver of another vehicle who happens into the path of the vehicle in automated mode? What assumptions can be made about the AV stopping or taking evasive action, without the other party being determined as 'contributory negligent'??

There have been several well publicised (often fatal) accidents, mainly in the United States, involving a Tesla car on autopilot and a stationary police car, stationary fire trucks, and stationary white sided trucks, which raise further questions about whether drivers can assume the capability of the AV mode of operation, and how much the autopilot can be relied on, and so not paying as much attention as if they were driving themselves; and even if paying reasonable attention, whether or not they could react quickly enough to avert the crash.

The situation where the vehicle is in AV mode, but with the driver monitoring, raises interesting legal issues that need to be defined.

Was it the driver/monitor or the vehicle itself who makes a decision that leads to an accident?? Who or what is to blame? How do you establish if the vehicle was being properly monitored when an accident occurred? Was the human driver's reaction appropriate in taking control at a given time? How long is an appropriate reaction time when in an AV fall-back to driver situation?

Without resolution of these, and other, issues, it is likely that the take up of the vehicles themselves will be slow as all parties – manufacturers, insurers, users err on the side of caution.

A current UK Law Commission review is examining legal obstacles to the widespread introduction of AVs and will highlight the need for regulatory reform. While the Law Commission review will go some way to dealing with the points raised in this chapter, the laws surrounding automated and semi-automated vehicles appear very inadequate in their current form.

Within the context of the EU, another aspect is the Driving Licence Directive 2006/126/EC (n.d.), which should be amended to include specific training and licencing on DAS and ADS automation and how to use the technology – including disengaging and re-engaging.

But at the EU level there is a lack of harmonisation of the rules on liability in case of damages caused by accidents involving motor vehicles, but, under 'subsidiarity' Member States are responsible for rules on liability in their own state, with the consequence that there are different liability regimes across EU Member States.

The Motor Insurance Directive 2009/103/EC (n.d.) needs be revised in light of the need to clarify liability for both a DAS and ADS vehicle. Directive 2009/103/EC (n.d.) relates to insurance against civil liability in respect of the use of motor vehicles, and the enforcement of the obligation to insure against such liability. This directive enforces all vehicles in the EU to be insured against third-party liability and establishes minimum thresholds for personal injury and property damage coverage. In view of the approaching vehicle autonomy, an insurance on manufacturers' liabilities may be required.

Product liability for defective products under Directive 85/374/EEC (2019) is also relevant. This Directive considers that manufacturers can be held liable for any damage caused by a defect in their product. In case of an accident, either the driver or the manufacturer or both of them may be considered liable by a judge, depending on the exact circumstances in which it takes place.

In Directive 85/374/EEC (2019) the concept of 'strict liability' appears, meaning liability in case of no fault by any party. It gives a legal basis for those situations where no one is held liable, neither the driver nor the manufacturer, so that traffic victims are compensated. The liability issue raises important concerns.

Yet, there is currently no framework in place for harmonising the rules on liability for damages caused by collisions in which motor vehicles are involved.

In the United States, NHSTA has identified three main recommendations:

- A need for further clarifications,
- A need to ensure that confidential data is protected, and
- A need to keep the safety assessment letter's regulations as simple as possible.

The 'Self-Driving Coalition for Safer Streets' (coalition of automotive manufacturers and technology firms) has requested that U.S. regulators change federal automotive safety standards that effectively prohibit the operation of a car without the presence of a driver (Beene 2016).

Lead participants are Ford Motor Co., Google, Volvo, Uber and Lyft, and the request the NHTSA and Congress to revise federal motor vehicle safety standards prohibiting a fully automated vehicle.

Most liability rules are based on the concept of causality of the accident to determine who is held liable. With DAS and ADS it will be increasingly complicated to identify the exact cause of an accident.

The use of EDR to determine the exact circumstances of an accident and the deriving liabilities is being widely discussed (including within UNECE WP29).

EDRs are now required to be fitted in new vehicles in the United States, and, despite strict privacy rules in the EU, it is expected that EDRs will soon be fitted into the vehicle during its production, and will be covered in the Directive 2007/46/EC on vehicle approval.

Automated driving will change the vehicle insurance paradigm. Governments will hold insurers initially liable to recompense the insured, and transfer the onus from the insured to let the insurer determine liability (by negotiation or in court) with the manufacturer. This therefore transfers some liability from vehicle owners/drivers to vehicle and/or systems manufacturers. Liability laws are an aspect of utmost importance in this new paradigm and its perception will directly influence public confidence about AV. As vehicles become more complex with highly automated systems and connectivity, liability assignment also gets more complicated.

Liability issues are also under review within the context of the European Commission's Digital Market Strategy.

GEAR 2030 Working Group 2 on automated and connected vehicles, has discussed the use of data storage for liability purposes (GEAR 2030, 2016). Its opinion is that data storage will need to become mandatory at some point to establish whether the driver and/or the AV are in charge of vehicle control when an accident occurs. It recommends that a set of specific requirements need to be developed, and that data storage should be part of the Type Approval Regulatory framework and should deal with a number of related aspects: data integrity, data privacy and cybersecurity. This is now being worked on in UNECE WP29.

The issue of how police enforcement work will change with the rise of fully automated vehicles, for example, determining who was to blame in case of a collision.

8.24   Level 5 May Take a Long Time to Instantiate

The path to a world where most vehicle movements are automated are probably more than 25 years, possibly 50 or more years away. In the meantime, the reality will be ADAS in the 2020s, with increasing amounts of control. Level 4 systems are unlikely to become widespread till the 2030s, and Level 5 maybe another decade or more behind that.

Chapter 11 makes forecasts about the likely timescale of implementation and spread of AVs.

But having read this far the reader will probably already gathered that J3016 Level 4 and 5 AVs on mixed-traffic roads are much further away than advocates and some vehicle manufacturers would have you believe. That will also seriously affect the use of AVs in the MaaS paradigm, which cannot be an expectation in the near term.

How do you deal with these issues.? Well, attractive though it is made to sound, the first advice is 'don't believe the hype!'.

Local authorities and road operators should use this breathing space to plan and adapt their infrastructure for the onset of AVs, because they are surely coming, but can only work successfully if the infrastructure is ready to receive them.

9

Potential Solutions to Overcoming Barriers to MaaS

9.1 Addressing General Issues

Mobility as a Service (MaaS) has its place, but it will not take over entirely. Urban futurists like to dream of a world where citizens 'optimise' their transportation, to optimise its efficiency, but in practice citizens 'optimise' their transport to maximise their own comfort and their own efficiency, not that of the city. Developers of MaaS service provision need to take this into account if they are to stand a chance of a successful business proposition.

We also have to consider MaaS in two paradigms – MaaS today with existing transport modes (including driver assisted ride hailing) – and MaaS of tomorrow with driverless ride hailing. The reason this is a significant difference (when the service offered is essentially the same taxi service) is that the cost viability changes significantly. Whereas driver assisted ride hailing is a relatively expensive option that limits its take-up, it is predicted that the lower-cost driverless version will change the cost perception to a point that many drivers will elect to give up vehicle ownership in favour of this new mode. (This, however, assumes that the choice to travel by an owned car as opposed to a shared car is price elastic. That is to say the user will make his/her choice dependent mainly on price, whereas we have shown in the above chapters that in practice the choice to use an owned car is quite inelastic, and price plays only a relatively small role in this choice).

The impact of moving from a product model (buying then using a car) to a service model, where users 'rent' a service and are likely to pay for certain services on a sustained basis, e.g. software system updates, means that there are likely to be major changes to the business proposition. Smith (2014) predicts 'The automated vehicles of the future may be co-piloted by companies as much as they are by computers'. Seba (2016) predicts that 'Cars as a service could offer the same level of service than the one offered by car ownership but at a much reduced cost, namely 10 times cheaper. The asset utilisation will increase manifold from a roughly 4% (as vehicles are parked 96% of the time) to around 90% of the time'. These predictions are probably an overstatement, but the most significant cost of running a taxi service is undoubtedly the cost of the driver, so the change will be disruptive. However, it can only happen when and where Level 5 automated vehicles become practicable, so, as we have seen from other chapters, is not likely to happen near term, except perhaps in some limited urban contexts.

In the past, in many countries, pressure has been to reduce taxation, and therefore, minimise the cost of public transport provision. But in view of pressure to reduce pollution in cities, and political pressure to provide car-free liveable spaces in cities, and

Automated Vehicles and MaaS: Removing the Barriers, First Edition. Bob Williams.
© 2021 John Wiley & Sons Ltd. Published 2021 by John Wiley & Sons Ltd.

pressure in the attempt to curb global warming, it is likely that there will be a growing trend for governments to intervene, both by increasing subsidies for public transport (primarily to obtain a greener, cleaner, higher quality public transport service), but it is also expected that in many places city authorities may choose to operate MaaS service provision as a public service for their city, as an incentive to reduce car usage in the city. Either funding a specialist operator to run the service for the city, or establishing its own organisation to do so. Already many cities are involved in EU-funded projects to test out/trial MaaS service provision.

However, the focus of most MaaS trials are city centres, but the real challenge for MaaS will probably be in the suburbs. We saw in Chapter 7, Section 7.3 (level of social engineering readiness) that in many countries the spread of suburbia and outlying commuting towns has been encouraged over more than a century. Many, indeed most, of these areas are already poorly served by public transport. If driverless ride hailing services are introduced at competitive cost, and is practicable, this will likely take traffic away from public transport, adversely affecting its already fragile viability. But adverse effects do not stop there. Because of the distances between pick-up points, there would be a significant increase in kilometres travelled. Sivak and Schoettle (2015) estimate an increased travel per vehicle of up to 75%, (MacKenzie et al. [2014] as cited in LaMondia et al. (2016)), estimate that it 'could result in an increase in light duty vehicle travel between 30% and 160%.' Further, the distances travelled in a day by a vehicle could make the use of electric power questionable (although it is expected that that problem will be solved before AVs hit the streets).

So while MaaS will offer an optimum mix of modes for travel in a city, there is still a question as to whether it provides innovation/efficiency in the suburban setting, or is counterproductive. For example, London and the 'home counties' has a population of about 15 million, of whom only 3.3 million of these (22%) live within 'Central' London. While MaaS will clearly benefit the 3.3 million, the residual 12 million may not benefit, or may benefit but exacerbate the congestion problems.

This then raises the issue of whether MaaS is providing simply a social service, or whether it is a tool for social engineering. Certainly the European Commission, governments and local governments are keen to use MaaS to socially engineer a change in the way people travel.

9.2 Essentials to Enable MaaS

From the experience of recent and ongoing European MaaS projects and trials it is clear that the following issues are crucial enablers for MaaS to be able to succeed:

- Trust
- Impartiality
- Cooperation
- Integration services
- Commercial agreements
- Solid governance model
- Data protection

9.2.1   Trust

MaaS service providers are often in competition with each other, and certainly see themselves in competition with each other. For example, a bicycle hire business is in direct competition with an electric scooter ride service, and both see themselves in direct competition with a local bus or tram service. Public transport sees itself in direct competition to ride-hailing services, and many PT operators see the advent of driverless ride-hailing as potentially disastrous for them.

Yet, in a MaaS paradigm, these competing services have to share information (often perceived as sensitive information, such as transport running behind schedule, or unavailability of cycles/scooters at particular locations). This makes it very difficult for the involved actors to trust any of the other actors to be the service broker (e.g. for the local public transport operator to run a MaaS service provider function). The most successful brokers to date are either strongly commercially oriented service providers (such as Whim [https://whimapp.com/uk], etc), or consortia jointly owned by the participating actors.

It is clear that, without trust, a MaaS broker service will not succeed.

9.2.2   Impartiality

This issue is related to that of trust, but it affects both service providers and service users. If the service provider cannot trust the broker to provide an impartial and unbiased service, then trust breaks down.

Similarly, the service user must also believe that the broker is providing an impartial and unbiased service, and is not making the suggested journey proposal to maximise the broker's income, or trust is lost in the broker, indeed the whole service.

It therefore goes without saying that agreements between service providers should not include, indeed should prohibit, commissions based on allocated and purchased journeys. This in itself poses an organisational issue, because the size of service providers businesses will vary greatly and so the funding structure will need to reflect this. Some relationship of fees to organisation size will usually be necessary. This can be arranged on organisation turnover, but then the decision-making structure should not be weighted on organisation size, otherwise decisions may be weighted in favour of the largest, and that may affect route preferences. These matters will have to be openly discussed at the outset, and clear policy decisions made to both provide impartiality, and also to demonstrate impartiality to the public/service users.

9.2.3   Cooperation

This issue is also related to that of trust. Cooperation between service providers is one of the most critical, and difficult, issues for MaaS, because they are often in competition with each other. It is difficult to find resolution to this issue, but resolution is essential.

As we have seen in Chapters 5 and 7, and in Section 9.2.1, MaaS is a service that is made up by the provision of a mix of (often competing) services. It is clear therefore that cooperation and coordination between the services is required, and that, for the reasons stated in Chapter 7, actors may be reluctant to cooperate, and, more particularly, share confidential information with competitors.

Here, the problem is exacerbated if one of the actors (most likely the local public transport operator) offers the 'broker' service/hosts the MaaS 'app'. Early projects considered the local PT operator to be the obvious host, but current thinking tends to favour an independent 'broker' service with strong privacy agreements between the 'broker' and service providers. Independent service providers also have the benefit that they are more fleet of foot in developing and upgrading software and the 'app', but the disadvantage is that it is likely to be a more expensive solution (as some subsidised PTs offer as a 'public service' either funded, or subsidised).

While the direct beneficiary of the provided MaaS service is the traveller, given the political/social objectives for MaaS, it may often be appropriate for a local administration to consider a MaaS broker service as a local social service, and fund the service (operated separately from the local PT provider), with clear privacy arrangements in contractual agreements between service providers and the broker.

Even this does not entirely remove the issue, because comparison of the performance data between transport options is a clear component in proposing a MaaS journey. So, the provider of the less advantageous option will question whether or not to participate with others, or whether it would be better to market its services alone. Where there is a clear pricing differential (the taxi is dearer than the bus, but more convenient and faster) declared user preferences can make the choice. But where you have, for example, competing bus services, or a PT tram competing with a commercial bus service, this issue is present.

Where an authority wishes to politically move to MaaS, it may consider requiring participation in the local MaaS broker service to be a condition of licence approval.

9.2.4 Integration Services

Integration requirements present themselves at several different levels in the MaaS paradigm.

Firstly, the broker needs to receive data in a useable and unambiguous form. Requiring data in a specific form may sound simple, but for an established, large, and complex public transport authority, or commercial bus operator with legacy information systems, could find this difficult or expensive to achieve, and this presents a challenge. In practice the broker will have to take what he can get from the sources, and make the conversion to a standard format.

Moving to the integration of the journey, there are both technical and commercial issues to solve.

As we saw in Chapter 7, different operators use different systems, so generating a 'ticket', that can be used on multiple transport means, creates a problem. This problem incurs major expense to solve, and the practical short-term solution may be simply to abandon the idea from the scope of the instantiation.

Moving to accept credit cards as the payment means is probably the lowest cost solution to find a ticketing medium that can work for all transport modes and is probably the most practicable solution, but still expensive to install for a transport company with a paper/card based system or other ticketing technology.

QR code on a cell phone presents an option, but is more difficult for micromobility.

Ticket integration also creates difficulties for different transport modes that have different pricing. This problem is in some cases insoluble.

A further (related) problem that has emerged in MaaS projects is that many public transport operators are unable to deal with post-payment. It may prove impossible, in many cases, to overcome this, the consequence of this is hard to accommodate.

9.2.5 Commercial Agreements

Many early projects were put together on a cooperative, goodwill, basis, but recent projects have found that it is essential to have clear commercial agreements at the outset, and note the experience that several have found these difficult to negotiate and agree.

Problems occur particularly where a single flexible routing ticket is envisaged, and service providers have different pricing models. Commercial agreements need to consider these issues at the outset, and limit ambition accordingly if needed.

Another issue to be dealt with in the commercial agreements is 'what happens in the event of service failure' and who is responsible for what aspects in the event the project is sued.

9.2.6 Data Protection

Data protection in this respect has two aspects:

a) Cybersecurity
b) Data protection

9.2.6.1   Cybersecurity

The 'app' and the broker operating systems will be subject to the usual network and over the air risks. In Europe these systems will have to comply to Directive (EU) 2016/1148 of the European Parliament and of the Council of 6 July 2016 'Concerning measures for a high common level of security of network and information systems across the Union'. Elsewhere in the world, similar legislation will apply.

The risks are serious, but are no different for MaaS than any other OTA/ internet and networked systems. This is not a book dedicated to online security, and there are many works on that subject. So, suffice to say that MaaS brokers and any other connected MaaS systems need to be cybersecurity aware and keep their protection up to date.

9.2.6.2   Data Protection

Within Europe, the requirements for data privacy and data protection are probably the strongest around the globe, and are enshrined in an EU Regulation, commonly known as General Data Protection Regulation (GDPR).

Regulation (EU) 2016/679 is the EU law on data protection and privacy in the European Union (EU) and the European Economic Area (EEA).

2016/679 also addresses the transfer of personal data outside the EU and EEA areas.

This Regulation is designed to give control to individuals over their personal data and to simplify the regulatory environment for international business by unifying the regulation within the EU.

2016/679 supersedes and replaces the previous Regulation (the Data Protection Directive 95/46/EC)

2016/679 contains provisions and requirements related to the processing of personal data of individuals (formally called data subjects).

Any MaaS system that acquires personal data has to abide by this regulation. MaaS instantiations and their 'apps' therefore need to be familiar with GDPR and abide by it.

MaaS instantiations also have to make their clients comfortable that their personal data is protected, that GDPR is being complied with, and it is recommended that this is highlighted as an important feature of the system. Failure to do this will be to the detriment of the project and seriously threaten its chances of success.

9.2.7 Solid Governance Model

9.2.7.1 Introduction
As stated above, many early projects were put together on a cooperative, goodwill basis, often without a formal governance model.

Project experience from many projects have stressed the need for a strong and clear governance model to be one of the first, and most important, objectives of any MaaS project/instantiation. Without this the project/instantiation is likely to fail.

The governance model should explicitly define risk ownership and elaborate clear policy and strategy for risk mitigation.

9.2.7.2 Governance for ITS Data Management and Access
MaaS is largely about using transport system and vehicle data from different sources. But automated vehicles and driver-assisted vehicles also need to exchange data with each other and with the infrastructure. But these are not the only intelligent transport systems (ITS) applications that need to share and transfer data. ITS are all about using, obtaining, and sharing data. Chapter 2, Section 2.7 shows the actors in a connected vehicle world (including MaaS), and also shows the connectivity of this paradigm.

But to date, as we have seen, data is in a silo somewhere and not available. Even worse is sometimes shared insecurely or is not adequately protected. There is little consistency about how it is handled, and in Chapter 8, Section 8.16 we discussed the need for cybersecurity.

And when you talk with experts on these subjects, they invariably ask – yes, but how is this going to be managed? And it is a very valid concern. For if it is not properly governed and managed, it won't happen. And it needs to be managed consistently. The US organisation 'The Auto Care Association' serves the supply chain of the automotive aftermarket in North America and elsewhere, and provides advocacy, educational, networking, technology, market intelligence and communications resources to serve the collective interests of its members. Their principal area of interest is neither MaaS, nor automated vehicles, but access to onboard data for maintenance and repair, and for fleet operations such as rental cars. But they powerfully argue the need for common security and common governance. They advocate a 'Secure Vehicle Interface' based on (existing) international standards and advocate a common governance model for ITS data management and access. Because the automotive market is at the same time global, but with regional differences, they propose a collaborative governance model (Figure 9.1).

While governance is more than government, the reality is that governments influence governance, and while UNECE WP29 and WP1 strive to develop common regulations, practices and agreements for the global road transport sector and encourage governments to become 'contracting parties' to these regulations, practices and agreements, there are regional and national differences.

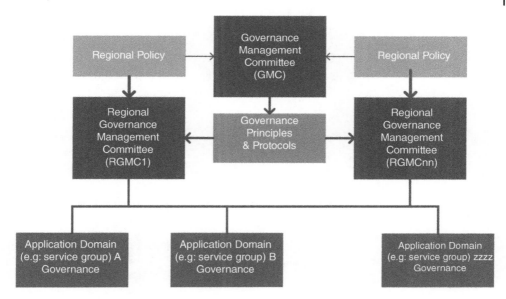

Figure 9.1 Conceptual governance reference architecture.

The Auto Care Association proposes that this governance specification assumes that there is the good will of one or more governments or regions, who, recognising and accepting regional differences, and recognising that vehicles are made and used on a global basis, will work together to establish and operate the governance management committee (GMC) for ITS data management and access.

The GMC will determine and agree overarching principals, protocols and regional variations.

For each region there will then be a regional governance management committee (RGMC) that will act as the governance executive for ITS data management and access for that region. It is assumed that the regional government of that region will either sponsor or determine how the regional GMC is funded.

The RGMC will implement the ITS data management and access for the region (in accordance with the GMC policies and regional government policies).

Each domain of ITS where ITS data is accessed and used will have its own application domain (e.g. service group) management administration process. (called a 'policy management committee' [PMC]).

Whether those application domains (e.g. service groups) are administered separately within each region, or are organised on a global sector basis (or covering several regions), shall be a matter of appropriate practice, but shall ensure that the requirements of the RGMC are met.

Looking at recent C-ITS cooperation between the United States and the European Commission, Figure 9.2 provides an example of how such governance might work in practice.

Here the GMC sets the common governance with input from both regions, and within the commonly agreed parameters, the regional GMC introduces local requirements and manages the day-to-day central administration (such as certificate management, etc.).

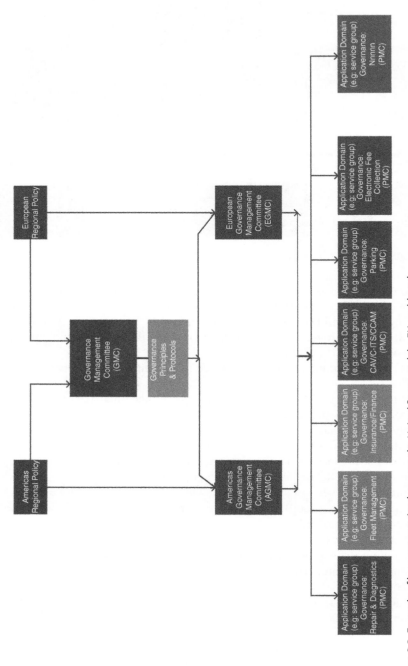

Figure 9.2 Example of how cooperation between the United States and the EU could work.

Where there would be separate application domain (e.g. service group) management, governance would depend on the structure at the application domain governance level. We can reasonably assume that in most cases, industry will press for one common PMC that will make necessary regional adjustments (as already in effect happens for C-ITS communications).

In the above example, six application domains (e.g. service groups) are envisaged. But the system is extensible to meet needs. Or indeed, in the reverse, two use-paradigms could operate a common governance administration (e.g. 'insurance' and 'fleet management'). One the other hand, where the use case paradigms are different (e.g. electronic fee collection) the US could run one application governance regime and Europe another.

Figure 9.3 (this is an example and not an implemented structure) shows a more extensive example where many governments, regional governments or associations of governments participate and may make input to the GMC, and manage their own regional requirements. It may start with two, (e.g. the European Union and the United States as shown in Figure 9.2). Canada may elect to join with the United States to form an 'Americas' RGMC. In Figure 9.3, Australasia is shown as operating a separate RGMC, but could elect to join with the European RGMC or the US RGMC. Other areas may establish their own RGMC at a later date, but all agree to work within the same high-level framework. The figure shows how, for example, the European Commission could establish a scheme on behalf of EU Member States, or APEC (or ASEAN) on behalf of APEC/ASEAN countries. This could operate on a regulated basis, a 'contracting party' approach through UNECE WP29, or a similar structure, or could work on a purely voluntary collaborative 'contracting party' basis.

The framework for governance, particularly while there is regional variation in policy and paradigm, is very abstract. So, if there is to be consistency, and interoperability at the implementation level, there has to be the use of common techniques and tools to achieve the required interoperability.

Adapting from the EU's Delegated act for C-ITS, and staying for the moment, at the level of 'roles', and without prescribing the nature of the regional organisational structure, the aspects that have to be performed can be described as shown in Figure 9.4.

And while the summary role of 'governance' is similar between the international agreement level, and regional governance level, the regional governance has also to design, manage, and operate control systems such as certificate issuing, operating registries, etc.

The process can be further analysed into the key roles that have to be fulfilled, and these key roles that need to be performed in order to achieve governance, are shown in Figure 9.5.

At the heart of the EU's delegated act, and the Auto Care Associations's Secure Vehicle Interface, lies a security credential management system (SCMS), (sometimes just called the 'credential management system' (CMS)), this is an established form of software that is used for issuing and managing credentials. SCMS exist for both symmetric and asymmetric cryptographic systems, with the public key infrastructure (PKI) being the most common for asymmetric key management. SCMS, and PKI in particular, encode the trust relationships and governance structures for identity and authority management, and thus lie at the technical heart of these governance specifications.

The SCMS of itself plays no part in the message security solution for vehicle-to-vehicle (V2V) and vehicle-to infrastructure (V2I) communication, rather it facilitates the

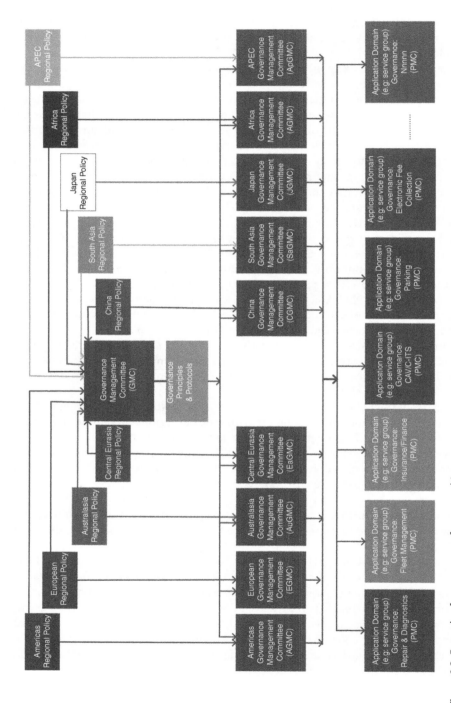

Figure 9.3 Example of governance reference architecture.

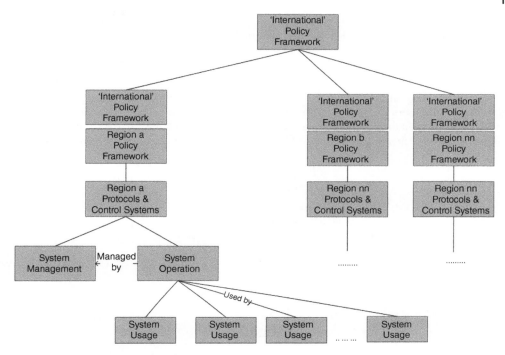

Figure 9.4 Organisational aspects (high level).

secure communication by ensuring that all participants in the system have access to the message specific, or application specific, cryptographic key material. The PKI approach assures public key certificate (PKC) management to facilitate trusted communication. A PKC carries the public key and an attestation by the issuing party that the public key is bound to an attribute of the holder of the matching private (secret) key. Authorised system participants use digital certificates issued by authorities within the SCMS to obtain public keys used to validate the content of signed messages. To protect privacy, these certificates contain no personal or equipment-identifying information, but serve as attribute certificates attesting that the holder is authorised to make the claim of the attribute (e.g. the specific value of an SSP) and that other users in the system can trust the attestation from the source of each message. The SCMS also plays a key function in protecting the integrity of the system by ensuring that keys and their associated certificates are revoked. The rules for revocation may include acting on reports of misbehaving devices.

Principal components of the SCMS are:

- A certificate authority (CA) that stores, issues and signs the digital certificates (note that CAs may be ordered in a hierarchy, with the top of the hierarchy termed the Root CA);
- A registration authority (RA), which verifies the identity of entities requesting their digital certificates to be stored at the CA (note that RAs are not mandated but may be used to relieve the processing burden on CAs);
- A key and certificate repository – i.e. a secure location in which keys are stored and indexed;

- A certificate management system managing things like the access to stored certificates or the delivery of the certificates to be issued Figure 9.6;

Note: In the European C-ITS paradigm (C-ITS delegated act and its annexes) they refer to their SCMS as the 'EU CCMS' [EU C-ITS SCMS].

Note: The European Commission's C-ITS delegated act and its annexes provide a good example of how to elaborate a C-ITS data management and access governance specification and its SCMS. While the scope of the EU C-ITS delegated act and its annexes is limited to C-ITS, and its implementation in Europe, and focussed towards one leading communications medium, it is governing

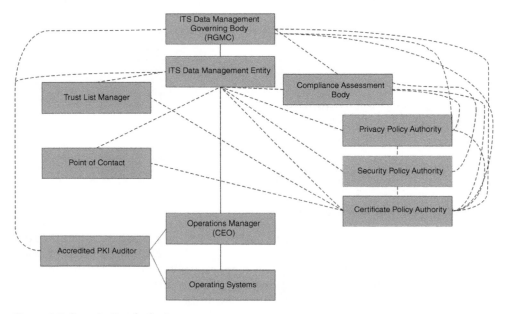

Figure 9.5 Organisational roles in governance.

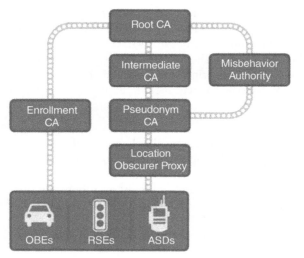

Figure 9.6 Simplified SCMS architecture. Source: US Department of Transportation.

the same paradigm of cybersecure wireless secure communications and data access between ITS entities, so provides a good example of how the GMC, and more particularly, the RGMC, needs to specify its procedures.

Both automated vehicles and MaaS will benefit and remove some of their barriers to implementation if a common governance structure, on the lines of that described above, becomes adopted.

9.3 Removing Obstacles to MaaS

Section 9.2 dealt with problems likely to be encountered, and ways to overcome them. This section describes obstacles to MaaS, that may be encountered, but there are no practicable solutions. They are just challenges that have to be tolerated, or borne by the project, and in some cases may terminally end any chance of success of the instantiation.

Public procurement is invariably a slow process, and ill-suited to a rapidly evolving paradigm. This just has to be tolerated, and taken into account in the aspirations and roll-out of the instantiation. Public–private agreements usually are cumbersome and slow to develop and obtain approvals.

Existing regulations will also provide problems (e.g. in UK electric scooters are not allowed on pavements or roads). Regulations for automated vehicles are still in a state of evolution, but restrictions, especially on early generation vehicles, may inhibit their use in a MaaS paradigm (in Level 3 automation 'disengage immediately on driver request'; Level 4 automation 'may issue a request to intervene').

Transport is a rapidly changing paradigm, and many of the changes are disruptive technology changes. There will therefore be technical obstacles, some foreseeable, others unforeseeable, that will have to be faced and overcome.

It should be noted that many public transport operators have reported that passenger claims – for everything from the consequence /cost of transport delays, back damage caused by inadequate vehicle suspension, clothing damages, injuries sustained on platforms and stairways, and many other reasons, reasonable and unreasonable, are already a major and expensive problem. This is within the paradigm of a single operator. A MaaS journey with multiple operators will infinitely complicate this problem, and who is responsible? Some of this can be managed through the governance model, but the problems will extend beyond that which the governance model can accommodate.

The infrastructure of the city, designed primarily around vehicle road use, or adapted to that use over the last century, is often not well suited to the MaaS paradigm. Adaptation will take political will, time and investment.

Additionally, reducing vehicle use in towns (a currently popular strategy) often has disadvantageous effects on travellers with limited mobility, because micromobility options, cycling, walking, etc., form part of the MaaS paradigm, and limited mobility travellers cannot use these transport means and often find public transport access difficult or impossible so have traditionally relied on road transport that can provide a problem-free door to door journey.

But by far the largest problem for MaaS is culture. Section 9.1 elaborates these issues and their background. But it is important to recognise in this section, that while there is political will for MaaS, its acceptability remains a question and may pose the largest problem for its use levels/instantiation.

9.4 Innovative Enablers for MaaS

The major enhancement that more recent MaaS projects have developed and evolved, is the role of the MaaS 'broker'. This is a new actor whose role is still evolving, but represents a significant enabling addition to the transport paradigm.

Some brokers are experimenting with individual mobility accounts that provide some assistance with ticketing in situations where the payment paradigms of transport operators are different. However, this seems to be being approached in quite fundamentally different ways in different projects, who recognise that a common standard will be needed.

10

The C-ART Innovation

10.1 Overview

As the above chapters have shown, quoted and referenced, there have been many research projects on the future of travelling. Most have limited themselves to consideration and development of a limited aspect of the travelling paradigm. However, the results of one project, C-ART (Coordinated Automated Road Transport) European Commission, Joint Research Centre (2017), initiated and managed by the Joint Research Centre, a Directorate of the European Commission, with the assistance of several sector experts, is worthy of detailed consideration, because it is provides the most far reaching study of where this may all lead to.

C-ART predicts that the future of driving/travelling will be radically different from what we know today, as a result of drastic changes that are expected with the introduction of automated and connected vehicles. It predicts that evolutionary and revolutionary scenarios may become real, with different socio-economic impacts.

Most significantly it predicts that, in order to optimise road journey movements, the new technologies, and particularly the advent of capable broadband wireless communications, some kind of central management would be beneficial, or may become necessary, in order to ensure that the full system benefits of road transport automation are obtained.

C-ART assesses research and predictions regarding the likely impacts of AVs on the transport networks, drawing on the work of Litman Anderson et al., 2016; Fagnant and Kockelman, 2015, Litman, 2016 and are qualitatively (by C-ART) summarised in Table 10.1 below.

10.2 Policy Context

C-ART proposed that traffic flow and system management in the presence of highly automated and connected vehicles, would be improved by central control. This is a fundamental proposal that is more radical than most other projections, but as this is a connected environment, and is largely data-centric, the efficiency of the system can be better achieved by control than random stochastic introduction/movement of vehicles.

The study presented a novel approach, based on a central coordination of fully automated and connected vehicles (i.e. C-ART), that policymakers and different

Automated Vehicles and MaaS: Removing the Barriers, First Edition. Bob Williams.
© 2021 John Wiley & Sons Ltd. Published 2021 by John Wiley & Sons Ltd.

Table 10.1 C-art qualitatively assessed.

Positive	Negative
Safety (↓ crashes due to human error)	Safety (↑ crashes due to new risk situations e.g. human factors issues in SAE level 3 systems, risk compensation, system failures)
Environment (↓ energy use / fuel consumption due to increased fuel efficiency and ↓ pollution due to reduced fuel consumption)[a]	Environment (↑ energy use / fuel consumption and ↑ pollution due to increased traffic)
Mobility (↓ congestion due to e.g. less delays that result from accidents, ↑ road capacity due to platooning, ↑ users e.g. young, elderly, disabled)	Mobility (↑ congestion due to increased travel demand, ↓ public transport)
Security (↓ criminal and terrorist activities thanks to vehicle control)	Security (↑ criminal and terrorist activities through hacking) and privacy (↑ risks of access to personal data)
Positive	Negative
Value of time (↑ leisure time) and comfort (↓ driver stress, possibility to rest or work)	Flexibility/joy/skills (↓ flexibility to take instantaneous journey decisions, ↓ joy of driving, ↓ driving skills)
Costs (↓ labour costs of taxis and commercial vehicles as drivers are no longer needed, ↓ crash costs if crashes are reduced, ↓ insurance costs if crashes are reduced, ↓ parking costs if cars can be parked in less space and located in less expensive land, ↓ car ownership costs)	Costs (↑ vehicle equipment costs, ↑ infrastructure equipment needed, ↑ maintenance costs) and revenues (↓ parking revenues for cities)
Business (↑ new business opportunities based on e.g. new mobility services, ↑ productivity)	Jobs (↓ jobs like taxi/truck/bus drivers and crash economy, ↓ vehicle repair demands if crash rates reduce)
Land use (↓ parking spaces and they can be located outside city centres, ↑ green spaces)	Land use (↑ sprawled development patterns as a result of lower Value of Travel Time)

a) further benefits with electric and lighter vehicles
Source: Own elaborations. (Project C-Art).

stakeholders may want to consider as a scenario for an eventual full realisation of a safe and efficient mobility system.

10.3   Key Conclusions

C-ART proposed advanced demand management strategies.

- Since C-ART relies on highly automated driving technologies (mainly Level 4 and Level 5), legal actions need to be undertaken to ensure that they can be safely deployed in real driving conditions (including without any occupant).
- Similarly, vehicle-to-everything (V2X) connectivity (mostly vehicle-to Infrastructure [V2I]) will be essential in C-ART as AVs need to communicate with the central management system.

- Road infrastructure requirements are indispensable in this context, mainly including communication equipment such as roadside units to communicate with AVs and with the management system, and digital infrastructure.
- Since C-ART requires that automated vehicles' algorithms are known by the system, (at least up to a certain degree), data sharing becomes an essential pillar. This is linked to crucial aspects such as data privacy and data security. However, is not fundamentally different from a vehicle providing its journey to a sat-nav map provider in order to optimise routing taking into account dynamic traffic conditions.
- The use of both dedicated in-vehicle interfaces for C-ART relevant messages and external vehicle HMI to inform about vehicle's intentions to other road users are understood to be of relevance for an appropriate user experience with the C-ART system (thus contributing to users' acceptance and in the end adoption of the system).
- Proper care should be given in the definition of rules and criteria governing the CART real-time decision-making process.

C-ART suggests that, in the long term, automation could have a revolutionary impact on travel behaviour and urban development. C-ART proposes that it is logical that the automated driving paradigm, where vehicles are able to move without human intervention (automated), they should also be coordinated in order to maximise the overall efficiency of the transport system (connected and coordinated).

C-ART proposes that a vehicle's connectivity will be the first element, automation will follow, and it will represent the final enabler for the management of the entire transport system.

However, the C-ART project noted that the road transport system is complex and the potential impacts of these technologies, which could contribute to totally reshape the vehicle use paradigm, are mostly uncertain and could even have undesirable consequences.

In particular, C-ART presented two main scenarios based on the expected socio-economic impacts of automated and connected vehicles in two timeframes:

- C-ART 2030: AVs account for a small percentage of vehicle fleet and co-exist with conventionally driven vehicles. Big safety risks remain. Road capacity is *reduced* in this scenario. C-ART could provide partial benefits in such a scenario, as it will first be implemented in parts of the whole road transport network. For vehicles travelling through the C-ART network, reductions in accidents, congestion, fuel consumption and emissions would be expected.
- C-ART 2050: AVs represent a substantial portion of road transport, almost achieving a 100% penetration rate. Road transport demand increases significantly in this scenario. Road capacity is also increased but, at times, may not off-set the effect of increased demand in certain points of the network. Full system benefits can only be guaranteed when C-ART controls the whole road transport network. As consequence, AVs would smoothly travel, minimising the environmental impact while providing very high safety and comfort levels to end users.

These scenarios highlight the potential need and benefits of implementing a central coordination system and serve as a basis for a preliminary conceptual definition of the CART system. A number of C-ART framework requirements are elaborated in the C-ART report, identifying where EU action could be of special relevance. Technology, policy and users are three main pillars to be taken into consideration in the transition to fully automated and connected vehicles, eventually enabling C-ART.

10.4 C-ART Scenarios

C-ART proposed two scenarios::

10.4.1 Short- to Medium-Term Scenario (2020–2030): C-ART 2030

C-ART provides coordination of AVs in some parts of the road transport network (e.g. on specific highways). AV users would thus need to decide whether they want to access the C-ART network at a given point in time. In this decision-making process, users could rely on updated information about the status of the C-ART network, showing, e.g. estimated entry time, estimated travel duration, estimated fuel consumption and emissions reduction. If users accept to drive through the C-ART network, they will receive an assigned time slot for network access, which will depend on the current use demand. They will be navigated to the access point of the C-ART road to be used. This scenario may require the existence of specific infrastructure areas at entry points to facilitate the access of AVs to the network. When leaving the C-ART network, passengers in AVs or vehicle owner/manager will receive a message with real consumption data, showing how C-ART has contributed to a more sustainable mobility.

Expected C-ART impacts: A C-ART system could provide partial benefits in such a scenario, as it will be possible to implement it in just a part of the whole road transport network. For vehicles travelling through the C-ART network, reductions in fuel consumption and emissions would be expected.

10.4.2 Medium- to Long-Term Scenario (2030–2050): C-ART 2050

AVs represent a substantial portion of road transport, almost achieving a 100% penetration rate. Travel demand increases significantly in this scenario. Road capacity is also increased but cannot satisfy the increased demand in certain points of the network.

C-ART will coordinate AVs along their complete journeys. No decision making will thus be needed on the side of the users. The C-ART system will manage the existing demand at a given time and will thus allocate AVs to different routes, optimising safety, fuel consumption and travel duration in real time. The use of fast and reliable algorithms for real time C-ART criteria calculation will make it possible to organise traffic in the best way possible, without penalising any user.

Expected C-ART impacts: Further benefits would be possible with this scenario, where C-ART would be able of controlling the whole road transport network. AVs would smoothly flow over the roads, minimising their environmental impact while providing safety and comfort to C-ART users.

10.4.3 Town Planning as a Consequence of C-ART

The design of towns and cities will change. At present these are laid out to facilitate (to a greater or lesser extent) easy parking. However, if vehicles are able to drop their occupants off and go away to self-park there could be significant implications for town and city planning.

It will be necessary to make sure that there is enough available parking reasonably close by, otherwise there is a risk that CAVs could drive themselves around towns after dropping off their occupants and cause an even more clogged-up road network.

This raises a further consideration for local authorities. At present they receive income from car parks they own and from fines for parking offences. The first of these could be reduced under an out of town parking site; however, there could be potential benefits from putting the land to alternative use. In theory, at least parking fines should become nonexistent under a system of full CAVs.

10.4.4   An Assessment of C-ART

In the opinion of this author, Project C-ART is the most forward looking, perceptive projection available to date, and one that considers the negative as well as positive aspects in its projections,

In criticism, it assumes acceptability and enforceability. History, however, teaches us that the option of individual control and user choice is a parameter that influences viability. Napoleon, Hitler and probably Genghis Khan, probably truly believed that their solutions were more efficient than those developed by those they domineered, but in the long term, the viability of their solutions only persist where they have public acceptability. Central *control*, as a directive, may not be acceptable, whereas, central guidance, could, if properly promoted, could prove popular.

Secondly, it also assumes, by 2050, dominance of C-ART, whereas the reality is far more likely to be a mixed system for the foreseeable future.

Thirdly, though they may seem long-sighted, as we have seen in the earlier chapters of this study, the technical ability to achieve 99% + safety of automated driving will not happen in the short timescales predicted by the car industry. C-ART's 2020–2050 scenario is more likely 2025/30 – 2060/70, and the post-2050 scenario is more likely post-2070.

However, the forecast of this report, elaborated in the next chapter, and based on voluntary, largely not mandatory, control, recognises the most significant contribution of Project C-ART.

10.4.5   Technology Principles and Architecture Behind C-ART

10.4.5.1   Research Origins

The TRAMAN21 (Traffic Management for the twenty-first century) TRAMAN (2019) project comprises five interconnected work packages (WP): (i) overview; (ii) traffic flow modelling; (iii) motorway traffic control (generic hierarchical control structure); (iv) local field test (v) dissemination. TRAMAN21 project, uses the terminology 'vehicle automation and communication systems (VACS)'. VACS are initially classified into the following three general categories:

VACS without direct traffic flow implications

VACS with traffic flow implications, which are further distinguished in:

- Urban traffic-related VACS
- Motorway traffic-related VACS

10.4.5.2 VACS Without Direct Traffic Flow Implications

The following category of C-ITS services is perceived to benefit from interaction with a central control system, but are not directly related to traffic flow management:

- Collision warning and avoidance systems
- Lane keeping assistance systems
- Vision assistance systems
- Speed monitoring systems
- Driver monitoring systems
- Other assistance systems
- Other warnings systems

10.4.5.3 VACS with Traffic Flow Implications

The project evaluated (by simulation) the effects on traffic flow where the VACS can control the vehicle routing, or mitigate the vehicle speed when approaching congestion.

Detail of their findings are somewhat dated so are not further discussed here, (but can be found at www.traman21.tuc.gr) but formed the start point for C-ART.

10.4.5.4 Lane Assignments for Autonomous Vehicles

It should be noted that a coordination system of AVs within a roadway has been patented by Amazon at the beginning of 2017 (US Patent US 9547986 B1) Curlander et al. (2017), under the name 'Lane assignments for autonomous vehicles'. The system generates lane configurations (e.g. travel direction, lane width, restrictions on types of vehicles) and roadway assignments (assigning, e.g. a lane, a time range and a speed range for access to the roadway) depending on roadway status and data, requests made by AVs, a roadway cost function that relies on different factors to calculate the costs to use a given roadway and an outcome directive that searches to optimise certain parameters (e.g. maximise traffic flow, maximise speed of vehicles, maximise toll revenues).

10.4.5.5 Development of C-ART

Using this background information, the C-ART system was described as an extension of the automated driving concept by adding communication capabilities that connect vehicles in between and with the infrastructure and adding a central coordination player that manages traffic on the basis of a set of criteria.

C-ART developers were all embracing. They postulated that criteria could go further than just routing and speed, but also take into consideration, e.g. fuel consumption and emissions, safety and travel time.

To be clear, C-ART is therefore founded on highly automated and connected vehicles (Level 5) and a connected infrastructure.

C-ART is presents itself as being the 'ideal' transport system that 'provides AVs with a central coordination and regulation in order to manage their access and use of the road transport system'.

It should be clear by now that C-ART is more than just automated and connected driving, with the infrastructure being the entire transport system, which should be in the position to instruct the vehicle on more fundamental choices to take (e.g. the path to follow or the speed to maintain). This is an important conceptual novelty with respect to the other paradigms.

C-ART proposes to take advantage of connected vehicles, particularly automated vehicles, communication and automation capabilities, a road transport management system (RTMS) can have the role to guide each vehicle through its entire journey with the objective to optimise the overall efficiency of the system.

It is clear that C-ART requires that all vehicles must be automated and connected and that the RTMS is able to simulate in real-time the movement of the AVs and their energy and fuel consumption as well as pollutant emissions in the case that these variables are included in the optimization of the system.

This then implies that vehicle logics and operations are known to the RTMS (at least to a sufficient extent), which C-ART estimates is not expected to be the case at least for the next decade (this author's opinion is that it might be two or more decades).

In addition, C-ART assumes that a RTMS has sufficient capabilities to manage the tactical behaviour of thousands, often hundreds of thousands of vehicles and to ensure that none of them will be excessively penalised by the optimization of the system. It is therefore clear that a C-ART system needs to be seen with a long-term perspective, as most of the conditions for its introduction will not be available before at least 2040 (this author's opinion is that it might be 2050 or later).

The central controller has the role of optimising a certain transport system to minimise a combination of the overall travel time and costs, energy use, air pollution and risk of accidents. It will be necessary to have reliable, though inexpensive models able to evaluate the status of the transport system, its short-term evolution and the connected externalities in real-time.

However, C-ART limits the role of the RTMS to road transport, whereas, its functions also put it in a close, but not the same, role to that of a MaaS service provider. To this author's opinion the RTMS could probably absorb also the role of MaaS service provider.

C-ART observes that traffic simulation models, fuel consumption and emissions models, pollutant dispersion models and collision risk models are necessary. Although many possible options in the different fields area available, an integrated solution going from traffic management to pollutant concentration and risk of accidents does not yet exist, and considerable work will have to be undertaken.

Once the model for simulating C-ART is available, different strategies for its optimization and for the localization of the critical situations can be adopted. Given the novel character of this activity, any result achieved on this point will provide a contribution on the state-of-the-art on AVs.

Two C-ART systems for the two projected scenarios:

- C-ART 2030: implemented on some roads
- C-ART 2050: implemented on the whole road transport network

10.4.6 The C-ART Framework

C-ART, of course, assumes C-ITS V2X technology/ Level 4/5 AV technologies, including a common cybersecure gateway (see Figure 10.1 below).

10.4.6.1 Telematics Architecture

C-ART summarises the telematics architecture as a composition of telecommunications and informatics that encompasses the computer and electronics in a car. Electronic

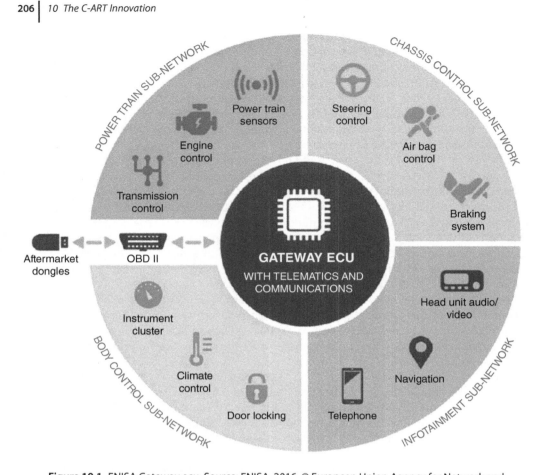

Figure 10.1 ENISA Gateway ecu. Source: ENISA, 2016. © European Union Agency for Network and Information Security (ENISA). https://www.enisa.europa.eu/

control units (ECUs) are micro-processing modules that form the core of the vehicle electronics system Lawson et al. (2015).

V2X connectivity (mostly V2I) will of course be essential in C-ART as automated vehicles need to communicate with the RTMS and could benefit from further communication possibilities.('Autonomous' vehicles, should such a beast ever exist, would not have the capability for C-ART.)

C-ART postulates that the AV algorithms would need to be known by the C-ART system, at least up to a certain degree and that this highlights the need for data sharing among relevant actors. However, no justification for this requirement is given.

C-ART poses questions concerning what data will need to be provided and maintained by road transport authorities? Which types of data will need management? What data will need to be stored, together with the related privacy concerns and security concerns, but it does not proffer any solutions.

C-ART expounds that the operation of such a system is effectively AVs connected to a central controller. It then poses questions for further research such as 'How should it optimize the transport system?', 'Who should govern it?', 'Prioritization / Optimization criteria?' Who determines the relative priorities between travel time, costs, energy use, air pollution, accident risk, etc? Is it the C-ART system, government, or are these driver choices?

Computational capacity and capability is assumed because of the rapid advances in computer technology and AI, but needs further study.

C-ART also asks the question as to whether AVs would need to undergo an examination to obtain a driving license or whether this could be covered through the type approval procedure?

Operational issues are also considered, both in general for AVs and the particular needs for C-ART, which would require significantly greater volumes of data exchange.

The project identified some basic infrastructure requirements for a C-ART system:

- Road infrastructure would need to be equipped with RSUs to communicate with AVs and with the RTMS.
- Road infrastructure would need to be equipped with traffic monitoring devices in order to monitor the driving situation.
- Road markings and traffic signs must be clearly visible at all times (although traffic signs will also be part of the map data).
- Digital infrastructure is of paramount importance for C-ART and is required to comply with high accuracy, frequent update rates, security, data protection, etc. Having a standardised data format is essential.

And raised a key remaining open question 'What are the specific data requirements for C-ART?'

In terms of message propagation, C-ART assumes that this will occur through V2V and V2I links. V2V links and can involve any vehicle nearby in order to build an end-to-end path towards a final point. C-ART identifies that one of the problems in V2V communication is that the vehicle network environment is dynamic and complex and sources of information can be heterogeneous.

Bou Farah et al. (2016) focused on the management of imperfect information exchanged between vehicles concerning events on the road. V2I links assume the presence of fixed road-side units everywhere in the infrastructure. C-ART references Campolo et al. (2015), who predict that a refinement of the existing technology standards in VANETs is required to support more advanced and complex use cases in a scenario with increased market penetration of equipped vehicles and road-side infrastructure coverage.

Security and privacy, two very important issues in VANETs, are considered critical in the development of robust VANET applications and C-ART opines that several network security issues resemble those of traditional wireless networks. According to Cunha et al. (2016), security challenges in VANETs are intrinsic and unique due to the size of the network, frequent topology changes, high mobility, and the different classes of applications and services, with conflicting requirements that will be offered to such networks. Integrity of the exchanged information as well as availability of the system are two main challenges regarding future generation VANETs, and C-ART systems.

C-ART observes that with regard to security and privacy, similarly to VANETs, Sun et al. (2015) highlights the fact that internet of vehicles systems, due to their characteristics of dynamic topological structures, huge network scale, nonuniform distribution of nodes, and mobile limitation, face various types of attacks, such as authentication and identification attacks, availability attacks, confidentiality attacks, routing attacks, data authenticity attacks, etc., which result in several challenging requirements in security and privacy.

In respect of human factors aspects of relevance for C-ART, the C-ART project identifies that the system would initially require:

- a dedicated in-vehicle interface that passengers can use for non-driving related activities would be convenient. In it, C-ART relevant informative messages could be included.
- an external vehicle HMI so that pedestrians, cyclists, powered two-wheelers (PTWs) can stay informed about the relevant vehicle intentions.

And the C-ART project notes that probably the most relevant aspect in the context of C-ART is users' acceptance and overall users experience with the system, as it will directly influence the real system use.

The C-ART project identified the following open questions are:

- How to manage a mix of AVs and conventional vehicles? (Problems arising from their interaction? Is retrofitting of old vehicles possible?)
- Should drivers have the right and freedom to overrule the controller's decisions? (Always, on certain time periods or in specific areas?)
- Willingness to pay for services?

However, C-ART while making quite detailed elaboration of the progress to automation, does not go as far as to make detailed proposals regarding the operational and functional architecture.

10.4.7 Some Observations on Project C-ART

The concept of C-ART is presented by the project as an evolution of the road transport management concept in the presence of connected and automated vehicles. C-ART would provide connected and automated vehicles with a central coordination and regulation in order to manage their access and use of the road transport system. C-ART proposes two timeframes that have been considered in this analysis, for which C-ART solutions have been proposed: a short-to-medium term timeframe (2020–2030) with CART in certain environments and a medium-to-long term timeframe (2030–2050) when CART could be expanded to the entire road transport system.

The C-ART project identifies a number of requirements have been derived for the proposed C-ART system, ranging from technology to infrastructure, human factors, data, ethics, insurance, liability, policy and legislation. Each of these areas highlight, as well, a number of open questions that would need to be further explored. Consequently, to face these anticipated challenges related to C-ART, the following main actions at EU level may be required:

- Develop a coherent EU regulatory framework that allows the deployment of AVs up to full automation together with V2X connectivity, including but not limited to traffic rules in the presence of AVs, driving licence, AVs certification, insurance and liability in AVs driving, connectivity, infrastructure, data privacy and security.

- Create a clear policy and legal framework for the data economy, enabling the free movement of data. A data-sharing and access policy in conjunction with standards in data formats would be necessary.
- Consider ways to improve public acceptance through demonstration activities or awareness campaigns.
- Analyse the specific policy and legal needs in relation to the road RTMS in C-ART.

Although a lot of uncertainty exists, the background data outlined in the project C-ART study suggests short- and long-term plausible scenarios where a C-ART solution could be beneficial.

While C-ART is incomplete (it describes itself as 'Stage 1') it sets the concept to progress the future development of the automated driving paradigm.

As previously noted, C-ART limits the role of the RTMS to road transport, whereas its functions also put it in a close, but not the same, role to that of a MaaS service provider. To this author's opinion the RTMS could probably absorb also the role of MaaS service provider.

C-ART also follows the assumption that fully automated driving will become endemic and indeed dominant, indeed ubiquitous. Given an environment where European drivers largely cannot give up control of the gear-stick, it is difficult to see them all giving up control of the vehicle. To this author's opinion, it is far more likely that there will be mixed fleets for the foreseeable future.

Further, and related, this is not just an issue of a desire to drive the vehicle (akin to the difference between manual shift and automatic vehicles). C-ART further assumes drivers will be prepared to be 'controlled' by the RTMS.

Guidance, as in route guidance, is widely accepted by drivers, but it is advisory. People are much more prepared to accept advice than control. This is not an issue for Level 5 vehicles, because the driver has already ceded complete control, and whether that control is determined by the vehicle software or the RTMS is academic. However, at Level 4, acceptance will probably be a problem. To be fair, C-ART is focussed on Level 5. However, it begs the question 'what is the marginal benefit of full control over guidance?'

C-ART is propounded on the belief that central control has efficiency benefits. But central guidance, if widely accepted can have similar benefits, and, in a Level 4 AV the vehicle occupant may see something that the RTMS is unaware of (accident, fallen tree, broken-down vehicle), and wish to either take back control or press 'alternative route'.

Taking all of these factors into account, Chapter 11 proposes a paradigm, the 'Managed Optimised Architecture for Transportation' (MOAT), and introduces a new actor, the 'Travel Optimisation Service' (TOS).

11

Potential Solutions to Instantiate AVs and MaaS: Managed Optimisation Architecture for Transportation (MOAT)

11.1 Managed Not Controlled

An important precept for a 'Managed Optimised Architecture for Transportation', and its differentiation from coordinated automated road transport (C-ART) is that it is a *managed* paradigm, not a *controlled* paradigm.

Earlier attempts at centrally controlled route guidance were not successful, largely because of the way that tests/trials were put together, and limitations on computing power/capabilities at the time. C-ART takes into account current and expected developments in computing, and further posits that road traffic can be optimised (in a cooperative-ITS connected and automated [CCAM] paradigm) only *if* it (traffic in cities) is centrally optimised and managed.

This author posits that that if optimisation is effective, and optimised routes offered on request, it will be used by most road vehicles – the widespread adoption and universal take up of both vehicle and personal route guidance has proved this.

Traffic flow depends on the volume of traffic. In light traffic, there is no congestion (except in the case of a traffic incident). In heavy traffic, information from the centre is already very important to minimise congestion, but congestion is minimised so long as most vehicles follow the optimised route. If a small minority of vehicles do not follow the optimised route it is unlikely to significantly affect traffic flow.

The valid point made by C-ART is that the unmanaged stochastic introduction of vehicles into traffic causes congestion, that could be minimised through central control. This author's contention is that if that guidance advice is offered, and is truly optimising, most vehicles will follow it. Therefore, if it is offered on an as incurred/requested basis, using the power of the central computing to optimise and provide effective optimised route guidance, the routing will be adopted by most vehicles, but without the centre having to manage and control every vehicle. This gains most of the benefits without the huge computing power (and cost) required to centrally control every vehicle. This concept forms the 'Managed Optimisation Architecture for Transportation' (MOAT) paradigm proposed herein.

MOAT offers, amongst others, the following advantages:

1. Simplicity
2. Flexibility of instantiation
3. Much lower computing requirements/costs
4. Retains privacy of the user

Automated Vehicles and MaaS: Removing the Barriers, First Edition. Bob Williams.
© 2021 John Wiley & Sons Ltd. Published 2021 by John Wiley & Sons Ltd.

5. User acceptability
6. Can be merged with MaaS travel optimisation/MaaS broker service provision

11.1.1 Simplicity

Central control of vehicles is possible with modern cloud-based computers, but requires massive computing resource and, importantly, reliable ubiquitous wireless communications. With the advent of high bandwidth 5G communications on the horizon, this is now theoretically feasible. But when, if ever, there will be adequate coverage for 5G, is questionable. A cooperative intelligent transport systems (C-ITS) application using the C-ITS communications link is probably the more preferred route, but further work needs to be done on the business case, as the computing costs would be significant.

By comparison, at its simplest, optimising a route based on current traffic centre information is a simple extension of existing route guidance software. This can be provided locally, or on a nationwide basis. This can be advanced by update exchanges with the vehicle at predetermined intervals. Such systems can therefore take the expected input (from route requests) of new vehicles into the system, and balance/optimise the traffic flow.

This is, therefore, by comparison, an infinitely simpler paradigm than full central control, but still offers the opportunity to better manage the stochastic inflow of vehicles into the system by providing (and updating at predetermined intervals) an optimised route to the vehicle.

11.1.2 Much Lower Computing Requirements/Costs

A consequence of a simpler system is a lower demand on computing resources and their associated costs.

11.1.3 Retains Privacy of the User

A significant advantage of MOAT is that it can be better designed to retain the privacy of the user (so long as the identity of the vehicle requesting the route is not retained).

11.1.4 Flexibility of Instantiation

In comparison to a full centrally controlled system, while the MOAT can be instantiated on a national basis, it can also be implemented on a town/city basis. This means that instantiation can be made locally, and cities do not have to wait for a national rollout of a hugely complicated system.

Also, it does not require all vehicles on the road to be equipped for this service. It can be installed in appropriate locations, and will operate only with equipped vehicles, but, even where this is a small proportion of vehicles, should improve traffic flow.

11.1.5 User Acceptability

Given reasonable privacy safeguards in its design, MOAT retains the privacy of the user. This would be difficult for a centrally controlled system. This is therefore not only easier to instantiate, but likely to be much more acceptable to users.

11.1.6   Can Be Merged with MaaS Travel Optimisation/MaaS Broker Service Provision

In Chapter 5, Section 5.4 describer the role of the MaaS broker, whose role is to optimise a traveller's journey, all or some of which may involve road transport. MOAT is all about optimising to a traveller's journey using one mode of transport. So the functions are essentially very similar to that of the MOAT route optimising service, and could be combined as a 'travel optimisation service' (TOS), which is more comprehensively described in Section 11.2.7.

Especially taking into account the social and political objectives for MaaS, described in Chapter 5, combining the role of TOS and MOAT could be very desirable.

11.2   High Level Actors in the MOAT Architecture

Note: For clarity, in this chapter a 'journey' is a single movement of one or more persons originating from a defined pickup location (pick-up point) with one or more pickup points and one or more drop off points to a final destination using one or more modes of transport.

11.2.1   Traveller Group (Traveller)

One or more than one person travelling on a single journey originating from a defined pickup location (pickup point) with one or more pickup points and one or more dropoff points.

11.2.2   Subscriber (Subscriber)

An actor (person or organisation) who subscribes to an offering from a travel service provider (TSP). A subscriber may or may not be one of the traveller group, but is responsible for booking the journey.

There may be multiple subscription models such as:

- Vehicle ownership linked to use of MOAT service
- On-demand hailing (similar to Uber, Lyft, etc.) with payment made per incurred journey
- Service provision plan, which is a recurring (e.g. monthly) subscription tailored around expected service (similar to mobile phone subscription), with a regular service charge and per event charges added to recurring subscription charge. (This may be provided by, for example, a car manufacturer, car rental company, travel agency, MaaS service provider, or some new dedicated entity.)
- Pre-subscribed charging (similar to TomTom, Garmin, etc.) subscribed at some pre-ordained event (e.g. purchase of an automated vehicle (AV), or other mobility device or sat-nav)
- Local or national government service by sign-up (free or paid)

11.2.3   Travel Service Provider (TSP)

Actor who arranges the travel movement for the subscriber, arranges and provides the travel means, and manages the journey to its destination(s). (This may be provided by, for example, a car manufacturer, car rental company, travel agency, MaaS service provider, ride hailing service, or some new dedicated entity).

11.2.4   AV Operator (AVO)

Actor that controls and operates the physical dynamic movement and is in physical control of (drives) an AV.

An AV operator (AVO) may be the TSP or an entity subcontracted by the TSP, or in some instantiations may be a function of the TOS. (An entity subcontracted by the TSP or TOS may, but will not necessarily be, the vehicle manufacturer, or a vehicle provision service [such as Uber, Lyft, Addison-Lee, etc.], or a combination thereof); or could be the owner of the vehicle.

11.2.5   Travel Information Provider (TIP)

Travel information provider (TIP) is a new actor, either national, regional, or local (or combination thereof) – actual instantiation at the determination of the nation state, which, as described in Chapter 4, and in Chapter 7, Section 7.3 makes regulatory data and dynamic data available as a 'unified traffic information system' via a 'data distribution interface' to map providers, TOSs, traffic management centres (TMCs), and potentially AVOs, as pre-trip and on-trip information.

11.2.6   Traffic Management Centre (TMC)

TMCs are existing actors who operate the control mechanisms to manage physical movement of traffic through the road network under their control.

11.2.7   Travel Optimisation Service (TOS)

Using data from the traffic information provider, the TMC(s) and public transport provider(s), the TOS is a (new) actor that dynamically coordinates and provides guidance/manages journeys through the road network, working in conjunction with TMCs, and may be separate from or an extension of a TMC. Its role is to register planned journey movements received from TSPs, optimise the journey according to dynamic traffic conditions and provide relevant regulatory and other data to the TSP/AVO.

TOSs are expected to instantiate as TOS – local – who manage the major volume of local movements; TOS – regional/national, who manage all longer journeys; and TOS – international, who coordinate long distance journeys, such as international haulage movements (possibly including platooning). The exact structure of TOSs is expected to be a decision of the national government. The funding model will be decided by the national government, but is likely to mirror the national model for funding TMCs.

In this MOAT paradigm, the TOS is usually the combination of vehicle route optimisation and MaaS broker.

11.3   MOAT from the Subscriber/User Perspective

Let us view how this will work, firstly from the viewpoint of the traveller.

Jim wants to make a journey with his wife and daughter to the IKEA superstore in a neighbouring town, on the way dropping his daughter off at school. Jim has several options to travel. He has the 'SupuDupaUber' app on his phone, he subscribes to Vodatravel on a monthly subscription, and he has his own AV in the garage.

On this occasion he speaks to Alexa, his virtual assistant, in his sitting room.

'Alexa. Plan journey, please'.

'Hello, Mr Smith. How many persons in your travel group today?'
'Three'
'How many pickup points?'
'One'
'The first pickup point is where'
'Home'
'I have that address as 1 The Avenue, Smalltown. Is that correct?'
'Yes'
'The first dropoff point is?'
'Elmtree School'
'I have that address registered. The second dropoff point is?'
'IKEA in BigTown'
'I have that location registered.
Are there any more dropoffs?'
'No'
'When do you want to set off?'
'9 a.m'.
You have three options: MaaS, Your AV, your red car. Which one today?
'What can MaaS offer?'
'MaaS option one is a shuttle to the train station, dropping off at Elmtree school on the way. There is a train at 09.32 arriving Bigtown Central at 09.55, and an IKEA automated shuttle to the store every five minutes from the west station entrance.
MaaS option two is an AV directly to IKEA. In current traffic conditions that will take about 27 minutes.'
'Option 2 please'
'Your own AV or your SupaDupaUber account?'
'SuperDupaUber'
'Your car is booked and expected to arrive at 08.58. Do *you want your start pass today to be facial recognition or fingerprint?'*
'Facial recognition, please'
'Arranged. Do you want me to book a return journey?'
'No thanks, I will do that by phone later'.
Have a good trip. Anything else I can do for you?
'No thanks. Bye'.
'Bye'.

He could of course have commanded the app to send the route instruction to his AV, or to the sat-nav in his manually driven vehicle. On this occasion two MaaS options were given, but there could be more multi-modal combinations (Figure 11.1).

11.4 MOAT from the Travel Service Provider Perspective

The TSP contracts with the subscriber (who pays any costs, charges and fees). The subscriber may or may not be in the traveller group.

The perspective of the TSP, and its modus operandi, will vary according to its business model, but at a high level can be described as:

- Operate user interface
- Receive request from subscriber
- Characterise request options
- Calculate viable travel options
- Confirm options to subscriber
- Receive subscriber selection
- Fulfil travel arrangements
- Provide confirmation to subscriber
- Monitor/manage progress of journey
- Acknowledge end of journey
- Process administration requirement
- Delete personal data unless subscriber requests favoured route saving

11.4.1 Operate User Interface (UI)

The TSP has to provide and operate the user interfaces (phone, internet, watch, car screen, etc.)

11.4.2 Receive Request from Subscriber

The TSP receives the journey request from the subscriber and determines the request using one of its user interfaces.

11.4.3 Characterise Request Options

The TSP characterises the journey detail (If necessary, by further interaction across the user interface, and/or from prearranged presets).

11.4.4 Calculate Viable Travel Options

Using preloaded data and dynamic data from relevant service providers, and any relevant information from the TOS, in some cases passing the journey request (anonymised) to the TOS for it to advise (or determine) routing, regulations, and restrictions. The TSP calculates one or a number of alternative journeys, including cost and arrival time.

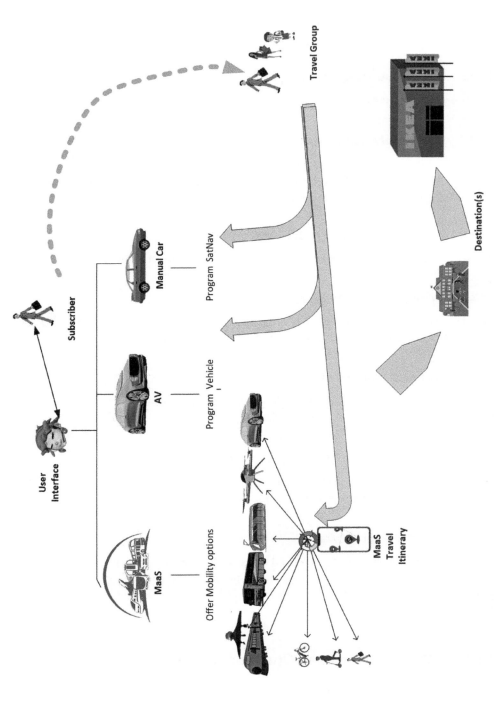

Figure 11.1 MOAT paradigm from the perspective of the subscriber and the user.

11.4.5 Confirm Options to Subscriber

The TSP advises the subscriber of the journey options via the user interface selected by the subscriber.

11.4.6 Receive Subscriber Selection

The subscriber selects a journey option.

11.4.7 Fulfil Travel Arrangements

The TSP fulfils the travel arrangements with the service provider(s) – such as instruct AVO, acquire tickets, make reservations, and if using MaaS, compile travel itinerary, make prepayments from the agreed payment account(s) of the subscriber.

11.4.8 Provide Confirmation to Subscriber

The TSP confirms the journey to the subscriber, and initiates the journey.

11.4.9 Monitor/Manage Progress of Journey

The TSP monitors the journey as it progresses through its phases, and manages any incidents or changes required in accordance with pre-agreements/trading terms.

11.4.10 Acknowledge End of Journey

The TSP acknowledges the end of the journey to the subscriber (who may or may not be in the travel group) by a pre-agreed means.

11.4.11 Process Administration Requirement

The TSP processes administrative paperwork, post-payment of subcontractors, TSPs, invoicing/payments, statements, etc., and closes the file.

11.4.12 Delete Personal Data

Unless the subscriber requests favoured route saving, or there are other contract provisions, the TSP anonymises records as far as possible, and generally complies with (in Europe) general data protection regulation (GDPR), or the national equivalent, privacy regulation.

11.5 MOAT from the Road Operator Perspective

The road operator (RO) (highways department, local authority, commercial operator, etc.) sees any road movements from the travel group in the same way as any other road user (be they on a public service vehicle, or automated vehicle).

In this MOAT paradigm, the road operator has equipped its road network with roadside ITS stations and is providing to and receiving data/information from equipped vehicles in the network (not all vehicles will be equipped).

The RO will have to manage a 'mixed' environment of equipped and not equipped vehicles, except where the RO dedicates roadways, or lanes, to operate in an AV only paradigm.

The RO will impose any UVAR (urban vehicle access restrictions) through control zone management (probably using CEN/TS 17380), and may well operate public service vehicle priority schemes using C-ITS station communications.

The RO may have lanes or even roads dedicated to AVs, and will manage those aspects through the transport optimisation service/TMCs in accordance with the local instantiation.

The RO is in close connection to the transport optimisation service and TMCs, indeed in some instantiations may operate either or both of them.

11.6 MOAT from the AV Operator (AVO) Perspective

An AVO may be a function operated by the TSP or an entity subcontracted[1] by the TSP, or may be a function of the TOS according to the local instantiation, which will differ from place to place.

The consequence of a range of options is that, within a cybersecure environment, the AV may receive its instructions (be operated by) different entities, and must be capable to accommodate and operate, either by direct instruction or forwarded instruction, in such paradigms.

In early generation instantiations, the AVO may be a person travelling in the AV who is charged with the responsibility to supervise/oversee the actions of the AV and surroundings and be ready to take over control from the AV if needed.

Whichever entity it is, the role of the AVO is the actor which controls and operates the physical dynamic movement and is in physical control of (drives) an AV.

The AVO receives its journey instructions, much as currently undertaken by a sat/nav map provider (who may indeed [but not necessarily] be subcontracted to be the AVO). The AV then controls the direction and speed of itself through its planned journey, having received, or receiving dynamically during the course of the journey, regulatory and other dynamic information from the TOS. Throughout the journey the AV will be in communication with other vehicles and the infrastructure/TOS via its ITS-station, and receive data/information from its sensors (such as radar, lidar, sonar, GNSS, gyroscopes, accelerometers, odometry and inertial measurement units) to sense its environment and perceive its surroundings, and use artificial intelligence to manage advanced control systems to interpret received information, in order to identify appropriate navigation paths, as well as obstacles, in order to move safely along its programmed journey.

The AV will advise the AVO of situations that it needs assistance to manage or optimise.

---

1 An entity subcontracted by the TSP or TOS may, but will not necessarily be, the vehicle manufacturer, or a vehicle provision service (such as Uber, Lyft, Addison-Lee, etc., or a combination thereof).

11.7 MOAT from the Travel Optimisation Service (TOS) Perspective

As stated above, the transport optimisation service is a (new) actor that dynamically coordinates and manages journeys through the road network, working in conjunction with TMCs, and may be separate from or an extension of a TMC. Its role is to register planned journey movements received from a TSP, optimise the journey according to dynamic traffic conditions and provide relevant regulatory and other data to the TSP/AVO.

The TOS provides service to connected and automated vehicles to optimise their routing, but also provides service to travellers to optimise their journey in a MaaS paradigm. That may be using a connected or automated vehicle for all or part of the journey, but is likely to involve optimising a multimodal journey, as described in the chapters above regarding MaaS.

The synergy between the TIP, MaaS broker, and vehicle route optimisation, lends itself to a common operation as the 'travel optimisation service'. But while in the C-ART paradigm this has to be an all-singing/all dancing powerful centralised function, in the MOAT paradigm there are multiple, and much less expensive, options, and greater flexibility regarding instantiation.

For example, most TOS requirements could be served by a city-based system, probably operated and funded in the same way that TMCs are funded. Whereas there could be a separate TOS to service commercial fleet requirements, including vehicle platooning management, probably operated by a separate commercial 'broker' service, and the same, a maybe a different organisation, providing long distance journey planning for light vehicles (may be a sat-nav company), which links into the local TOS encountered on the route.

The TOS gets static regulatory, and dynamic regulatory, information from the TIP, dynamic traffic information from the TMC, and fixed and dynamic public transport and alternative modes information from the MaaS broker, or be operating that service as well.

The TOS provides its service to any requesting traveller or vehicle.

While I have posited that a local TOS probably operated and funded in the same way that TMCs are funded, the exact structure of TOSs is expected to be a decision of the national government. The funding model will be decided by the national government, but is likely to mirror the national model for funding TMCs.

11.8 MOAT from the Traffic Management Centre (TMC) Perspective

TMCs are existing actors who operate the control mechanisms to manage physical movement of traffic through the road network under their control. The TOS will be a client and strong cooperation on dynamic traffic information will be required.

11.9 MOAT from the Travel Information Provider (TIP) Perspective

TIP is a new actor, which, as described in Chapter 4, and in Chapter 7, Section 7.3 makes regulatory data and dynamic data available as a 'unified traffic information system' via a 'data distribution interface' to map providers, TOSs, TMCs, and potentially AVOs, as pre-trip and on-trip information.

The TIP obtains its information from national government, local government/authority, in respect of static and dynamic regulations; timetables and dynamic operational data from transport service providers.

11.10 MOAT and Privacy

Clearly with any central system there is a risk to privacy, so MOAT systems need to adopt privacy by design. Central systems should use temporary identifiers, provide route guidance and remove the identifiers as soon as the vehicle has passed through the system and ended the journey.

11.11 The MOAT Overview Architecture

The MOAT overview architecture can be represented as follows (Figure 11.2):

In this diagram the MaaS broker service is shown as a separate actor, whereas it is logical to merge as part of the travel optimisation centre. However, for financial/business case/historical reasons this may be a separate entity.

11.12 The MOAT Systems Architecture

The MOAT overview architecture is of course an oversimplification. For example (Figure 11.3),

Is in instantiation (Figure 11.4):

We know from Chapter 5, that the simple box 'MaaS broker function' will in detail be something like (Figure 11.5):

And we have seen in Section 11.3 the MOAT paradigm from the perspective of the subscriber and the user.

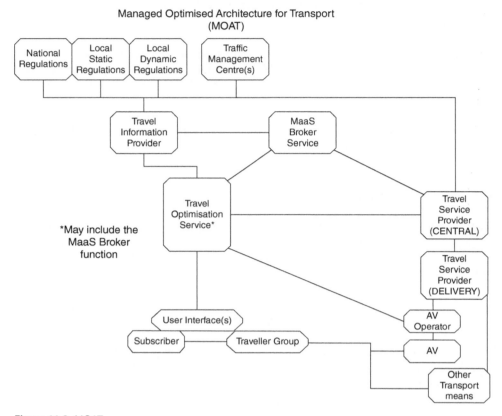

Figure 11.2 MOAT.

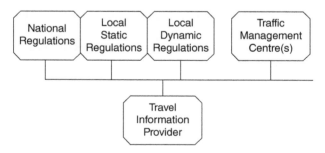

Figure 11.3 MOAT.

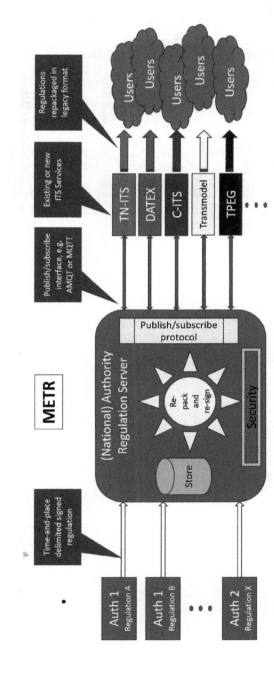

Figure 11.4 MOAT and management of electronic traffic regulations 'METR'.

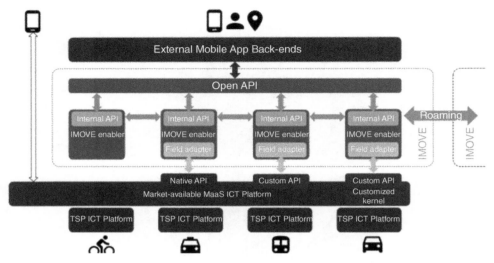

Figure 11.5 IMOVE.

12

The Business Case for MaaS

12.1 The Challenge

The problem in identifying the business case for Mobility as a Service (MaaS) is that it is proffered as a means to 'optimise' a journey. If is that simple, the business opportunity is to provide and operate the MaaS broker service, and everyone will be a winner. The traveller – the customer – gets his optimised journey; the travel service provider gets paid for providing the service; the broker gets paid for his service, usually by the travel service providers, but possibly indirectly by the traveller, probably by using the current funding model of financing the service by putting adverts into the 'app'. This paradigm posits the traveller as the 'customer'.

Now in city centres one can see a benefit to the customer – the traveller – being able to compare say the journey on the underground system, against the same journey by bus, against using a bike or electro-scooter, or an offered combination.

However, in most other situations MaaS is going to offer that the traveller travel from (a) to (b), wait for a connection, possibly in inclement conditions, to catch a link to (c) in order to wait for another connection possibly in inclement conditions, to catch a link to (d). In Chapter 9 we questioned why the traveller would elect to do this, particularly if his alternate option was to comfortably (in his own personal space) use his car from start point (a), seamlessly to endpoint (d) and back again.

The benefit of MaaS to the traveller will in these circumstances often be low unless the driving conditions are particularly bad, and even then, the advent of automated vehicles (AVs) will make difficult driving conditions more tolerable.

12.2 The Solution

The truth is of course that the real 'customer' is not the traveller, but the city authority on behalf of its citizens. This implies that it is the city authority who should fund the MaaS broker service, indeed, when taking all the operational problems described in Chapter 5, 7, and 9 into account, is probably the only viable business case for MaaS outside of the core of major cities where travellers are spoilt for choice of transport means.

Throughout this book I have noted the pressures to reduce pollution in cities, and political pressure to provide car-free liveable spaces in cities, and pressure in the attempt to curb global warming, that it is likely that there will be a growing trend for governments to intervene, and it should be expected (and a valid use of public money) that in many

Automated Vehicles and MaaS: Removing the Barriers, First Edition. Bob Williams.
© 2021 John Wiley & Sons Ltd. Published 2021 by John Wiley & Sons Ltd.

places city authorities may choose to operate MaaS service provision as a public service for their city, as an incentive to reduce car usage in the city; either funding a specialist operator to run the service for the city, or establishing its own organisation to do so. Already many cities are involved in EU-funded projects to test out/trial MaaS service provision.

12.3 The Outlook

We may expect these trends to continue, but even assuming that local authorities either directly, or via funding a 'transport optimisation service', fund the MaaS broker service. But it is unlikely that MaaS will take over as the dominant transport organiser, except in large city centres.

In the future, the use of AVs to provide the 'first mile'/'last mile' is touted as being a major incentive to use MaaS. In all probability, if AVs are available as cheaply as is being suggested, the availability of an AV will lead travellers away from MaaS, rather than direct them towards it.

MaaS will have its place, but it will primarily be in larger cities (where public transport is already the dominant mode of transport, so will not dramatically reduce private car use in the city). Outside that it will gain some modal shift, but for the reasons expounded in Chapters 5, 7, and 9, it is highly unlikely to gain the dominance that its promoters currently hope for.

13

The Business Case for Automated Vehicles

13.1 The Challenge

The outlook for automated vehicles is more certain. It is just the timescale to instantiation that is a challenge to predict.

The long-term business case is clear. One day a young girl will turn to her Daddy and say 'What do you mean Daddy, in the olden days people used to drive cars? Wasn't that very dangerous?'. According to the World Health Organization (WHO), road traffic injuries caused an estimated 1.35 million deaths worldwide in the year 2016 (2018); and WHO estimates that over 50 million are injured each year (2010).

But that little girl may be your grandchild, great grandchild, or even great, great grandchild.... Do not believe the predictions of Elon Musk, Ford, and Renault and other enthusiasts. If you are an adult today it will not be your daughter!

Chapter 14 will make some timescale predictions. So here is not the place to say anything other than it will take several generations before the automated vehicle (AV) is the ubiquitous mode of vehicle control.

What seems clear today is that the roll out of autonomous vehicle features will accelerate (lane-keeping assist, automated parking, adaptive cruise control, follow in slow moving congested lanes, automatic headlamp dip and road curve following, automatically change lanes, navigate autonomously on limited access freeways, and the ability to summon the car to and from a garage or parking spot, etc.), leading in a very few years (once everyone has finished arguing about which wireless communications technology is best) to 'connected' features where vehicles communicate with each other, and the infrastructure, to provide hazard advice and warnings, provide connected adaptive cruise control, merge and overtake assist, etc., and eventually, collision avoidance. At the same time, fully automated shuttles will (already are) providing first/last mile and shuttle services on strictly controlled, low speed environments, largely separated from normal road vehicle traffic.

Chapter 14 will speculate on the amount of time needed to make the final leap to fully automated vehicles and guess at how long it will take for AVs to become the ubiquitous form of transport.

13.2 The Solution

The business case solutions are there, but there are competing paradigms, and it is unclear as yet which will prevail, or which combinations will prevail.

Some, probably most, automotive manufacturers, currently see their role changing from being a hardware provider to a travel service provider, and encouraged by the McKinsey report *'Monetising Car Data- new service business opportunities to create new customer benefits'* (2016) appear to see relief from the oversupplied, huge turnover but low profit car manufacturing business, to a (greener grass on the other side of the fence) paradigm of selling data about their customers to third parties, and providing AV services to their existing customers, and to currently unserviced travellers, especially the young and old age groups and social classes too poor to currently own a car. The early chapters of this book exposed and detailed the weaknesses of this paradigm, and McKinsey failed to tell them that the taxi/private hire sector is even more competitive than car manufacture, and will become even more so when AVs are on the scene.

These same manufacturers also predicted a roll-out of AVs by 2020–2022. Well here we are in 2020, and as we have seen from the earlier chapters of this book, we are frankly nowhere near roll-out. It is not this year, it is not next year, but probably next decade, or even later than that.

There must be some very red faces at automotive makers board meetings at present. But after that embarrassment, what will be the succeeding business model for the OEMs?

If there are no significant AV and mobility as a service (MaaS) AV services for at least a decade, auto manufacturers will continue in the paradigm of the auto-manufacturer business case, and strive to increase profitability through increasing the electronics capabilities of their vehicles, for the next decade.

13.3 The Outlook

The next chapter makes the timescale predictions, but from having read your way this far through this book it must be clear to you by now that any such radical changes are not going to happen this decade, and possibly not the next, by which time we are well into governments – if they implement their current policies – requiring vehicles to be electrically powered, which produces just as much, if not more, change and business paradigm change opportunities/threats than the introduction of AVs.

In the meantime, vehicle manufacturers will continue to manufacture vehicles as their main business case, and find innovative financing schemes that rent the use of the vehicle to the driver, rather than sell its ownership, especially if they can use these schemes to acquire 'loyalty' (i.e. lock the buyer into their marque) at the time of vehicle replacement.

Data-based additional service provision (to the vehicle user and other third parties) will present some new opportunities, but fair and open market regulations, and privacy and data ownership regulations, will limit those opportunities severely.

The synergy between vehicle manufacturers and ride-hailing service providers will emerge as the popularity of ride-hailing over vehicle ownership increases (even though it will not take over), and ride-hailing providers will become the largest customers of the auto manufacturers, frequently leasing vehicle use from the auto manufacturers' finance divisions. So we may expect to see associations and mergers in these areas. For example, If Uber continues to struggle to make a profit, will they become vulnerable to a takeover by an OEM such as Mercedes, as a means to lock in their vehicle selection, or will a competitor do so to prize the taxi-drivers' current preference for Mercedes away to another Marque?

14

Timescales to Successful Implementation

14.1 Caveat

Comments have been made throughout this book regarding poor or ill-informed estimates of how long it will take to instantiate the paradigms that are the subject of this work. It is easy to criticise, and I have done so. So it is reasonable for the reader to expect that this author attempts to provide some better estimates.

This section attempts to do so. But must be read within the caveat of considerable uncertainty regarding the political and social issues and influences surrounding such forecasts. Part of my criticism of some of the forecasts cited in the early chapters of this book is their certainty. The timeframe to instantiation of these paradigms is very uncertain. It is preferable to understand that we are dealing with uncertainty, than to plunge headlong towards a certainty that will never come about.

I think that we can be reasonably clear that the remote control of the physical movement of vehicles, in terms of power on/off, acceleration, deceleration, and steering, etc., is largely a done deal. Like all new technologies, more efficient and better ways will emerge/evolve. But the major issues are solved. But that is the easy-peasy bit.

Sensing other vehicles, obstacles, discriminating between people, animals, and things like flying polythene bags, and even the seemingly simple task to identify large stationary vehicles, is not a done deal, as continuing accidents and deaths have proved. Maybe 90% there, but these are areas that need 99.99% efficiency. But there is some good brainpower at this, and from multiple sources, so I think it is reasonable to expect that these issues will be resolved in the next 2–5 years. It is implied that this will be completed by 2025.

Knowing the regulations – static and dynamic – is, however, nowhere near solved. Here we have two major problems.

The first and greatest problem is that most vehicle OEMs have thought, and continue to think, that their vehicle is at the centre of, indeed, is in control of, its movements through any traffic system. Whereas we have seen in Chapter 2 (especially Section 2.7), that they can only travel within the static and dynamic regulations applicable at that location and time, and within the dynamic instructions of the traffic management centre, who control traffic light operation etc. In this paradigm there are more than 20 actors, and the vehicle is not the controlling actor.

Until car designers/manufacturers understand this, accept and adjust their thinking to accommodate and operate within this multi-actor paradigm, it is highly unlikely that there will be a (Level 4 or 5) automated vehicle allowed on public roads in a mixed traffic situation.

Automated Vehicles and MaaS: Removing the Barriers, First Edition. Bob Williams.
© 2021 John Wiley & Sons Ltd. Published 2021 by John Wiley & Sons Ltd.

The second big problem is knowledge of the regulations at any place and time. This information has to be made digitally available to vehicles, for the reasons espoused and exampled in previous chapters. But again, as described in previous chapters, this has to be done by local authorities, who apart from Sweden, to some extent UK, and a few major cities, are totally unaware of this need, and even when they become aware of it, it will take many years to implement. But as Sweden has shown, there is just no way around this requirement. Cameras reading road signs is an inefficient, unworkable solution. How long will it take to get this realisation and take it through to implementation? My guess is for the most keen, advanced ad affluent cities, maybe 5 years. It took Sweden 10, and they were switched on. Think of the more backwards states of Europe and it could be 10–15 years, unless the EU provides funding, and even then will not be rapid.

14.2 Phased MOAT

One of the disadvantages of centralised, sweeping systems like those espoused by C-ART (Coordinated Automated Road Transport), is that they are so large and complex, that getting funding, actually designing and implementing the system, then correcting the unforeseen problems that arise, then work through generations of vehicles to get them equipped, will take so very long. A simple system like eCall took 16 years to implement, and then only into new model vehicles. Cooperative intelligent transport systems (C-ITS), whose architecture was basically designed by 2002 will probably get implanted into new model cars around 2022 or later.

The advantage of MOAT (Managed Optimised Architecture for Transport) is that it can be implemented piecemeal, as and where the local situation allows and implements it. True, the benefits are only local, but can, incrementally, immediately start to repay the investment.

The comparatively small central computing power behind MOAT could be instantiated on a national basis, but seem more likely that MOAT will be, at least at first, initiated locally, on a city basis. Because these local systems are limited, they can be introduced much more quickly. The catch, of course, is that these systems only work for connected vehicles/AVs, and that means the vehicles have to be equipped for C-ITS. Which largely (for anything other than simple Level 1 advisory systems), means being equipped by the OEMs during manufacture. Current arguments about whether this is achieved through the existing G5 5.9 GHz DSRC or the 5G C-V2X technology does not speed up such instantiation.

Another possibility for early generation systems would be via enhancing data to map providers to installed sat-nav systems (an upgrade of data from the travel information provider, to the sat-nav system provider, by introducing the supply of Management of Electronic Traffic Regulations [METR] [electronic regulations] [probably via TN-ITS]). This would provide travel optimisation (which is a process that sat-nav system providers already make best efforts to do), by improving the accuracy of the regulatory information that sat-nav systems currently use. Linkage between the travel optimisation service (TOS) and sat-nav provider (or subcontracting the operation to sat-nav service providers) by incorporating TIS/TOS information to vehicles connected only via their sat-nav system advising the driver, would bring many of the benefits sought by C-ART. However, this is an interim introductory step for conventional vehicles, and this book/chapter is about AVs and Mobility as a Service (MaaS). But it does, very importantly, mean that the investment in TIS/TOS does not have to wait for AVs to be on the roads to generate payback, but can bring many of the benefits to current generation vehicles and traffic management, and subsequently be available to AVs when they eventually arrive.

The critical feature is therefore the speed at which regulatory data (static and dynamic) can be made available to the TIS, and of course the speed at which TIS/TOS can be developed and instantiated. But the fact that many of these benefits can be brought to benefit current-generation vehicles, means that its payback is much more rapid and easier to justify.

14.3   Timescales MaaS

As stated above, we have to consider MaaS in two paradigms – MaaS today with existing transport modes (including driver assisted ride hailing) – and MaaS of tomorrow with driverless ride hailing using AVs.

MaaS with existing transport modes can be implemented immediately that the problems and obstacles described in the chapters above are solved – and these problems are commercial, organisational and political – not technical.

The biggest challenge that faces MaaS, at least in the short term, is the effect of Covid-19 on people's willingness to use public transport, ride-share, ride-hail, and shared micromobility. Will the trend to living in cities be lessened or even halted, thus increasing the need to purchase a vehicle, with its paid for availability lessening the attractiveness of MaaS? While Covid-19 is still unpreventable/uncurable, the whole concept of MaaS is dead. Assuming a vaccine, and/or effective treatment for Covid-19 arrives in the near term, how quickly will people feel more comfortable with public transport, ride sharing, and ride hailing? These are issues that only the passage of time will address. If Covid-19 is mastered and disappears from the scene, how quickly will the deep etched fears take to subside? Only time will tell.

MaaS including all or part of the service using AVs can obviously only be introduced when AVs are available, so the instantiation is only at the speed that AVs are instantiated (see Sections 14.4 and 14.5).

14.4   Timescales for Automated Vehicles

The provision of regulatory data from the TIS, and route optimisation from the TOS will be vital to the safety of AVs.

As has been commented above, the remote control of the physical movement of vehicles, is largely a done deal, it is 90+% complete, and should be a done deal before the end of 2025.

Volkswagen (VW) are equipping the upcoming Golf 8 with connected car capability, and others are following. The argument between 5.9 GHz and 5G C-V2X is slowing down the move to connected cars, but as seen from a recent presentation by Evensen (2020) shows that there is now a significant raft of organisations taking connected vehicles forward (Figure 14.1).

- Not quite...
  - http://c-its-deployment-group.eu/
  - Supported by 60+ Member States/OEMs/Cities/Operators/ Suppliers/...
  - All committed to deploy C-ITS according to DA/C-Roads/C2C-CC specifications
- Cybersecurity deployment continues
- CCAM continues the work of technical coordination, but aimed at next generation mobility and automation

Figure 14.1 C-ITS connected and automated vehicles (CCAM) deployment. Source: "Europe policy and projects". K.Evensen. MobilITS, Norway.

C-ITS Deployment Group is an association of vehicle manufacturers, road infrastructure providers, authorities and federal states, as well as industry and ICT and research partners, which aims to increase road safety and decrease congestions on Europe's motor- and expressways by deploying C-ITS, a technology to enable communication between vehicles and digital road infrastructure. It brings together the most prominent automotive manufacturing companies and interest groups in the traffic and transportation sector. http://c-its-deployment-group.eu.

Moving forwards, connected vehicles and low levels of automated vehicles (Levels 1–3) can move ahead quite rapidly once the wireless communication issue is resolved.

I have commented above (Section 14.1) that the safe introduction of AVs – Level 4 and Level 5 – requires the vehicles to be able to sense other vehicles, obstacles, discriminating between people, animals, and things like flying polythene bags, and even the seemingly simple task to identify large stationary vehicles is critical and needs 99.99% efficiency. I have made the prediction that I think it is reasonable to expect that these issues will be resolved in the next 2–5 years, which is implying by 2025, but be clear, we are not there yet.

But, as stated in Section 14.2, and in the earlier chapters of the book, trying to know the regulations of the road by reading road signs will never get more than 80–90% accuracy.

For an automated vehicle (AV) to be considered safe on the roads, in mixed traffic, it *must* operate within the regulations near 100% of the time. Otherwise an authority cannot allow it to operate without supervision. Reading road signs and getting it right 80%–95% of the time, and getting it wrong 5%–15% of the time, when the consequence will be at best a violation of the law, may cause a crash, and, at worst case, cause death or serious injury to the occupants of the vehicle, or third parties, is simply not acceptable. Map makers may hope and succeed to collate information over time from multiple vehicles passing any point, and that may overcome problems with signs being hidden by a high-sided vehicle, or snow, but it is unlikely to deal with contending, incorrect or incorrectly situated signs and markings. Neither the road authorities nor the designers of the AV, nor their insurers, can carry that legal liability and its consequences.

If you think that is exaggerating, go back and reread Chapter 8, Sections 8.6 and 8.23. Tesla, Uber, and others only escape punitive liability for dangerous design because in all current systems there is a watchful driver (theoretically) ready to take over at any moment of danger. A Level 4 or 5 AV does not have that luxury/excuse.

It is an unavoidable fact that AVs must have knowledge of and work within the regulations. Chapter 8, Section 8.3 provided ample evidence that this can never be achieved solely by reading road signs.

The critical feature is therefore the speed at which regulatory data (static and dynamic) can be made available to the TIS, and of course the speed at which TIS/TOS can be developed and instantiated. This is a 5–10 year process at best. And has the complication that OEMs have not yet understood/accepted the need for this yet and are still focussing on reading traffic signs, so it may take longer.

That means it is the critical path limitation to the introduction of the realisation of AVs. And it implies that AVs (apart from traffic separated/low-speed shuttles) can only be effectively operating at best by 2025, and probably not before 2030.

The following sections of this chapter provide what this author believes will be the probable timelines.

14.5 The First Half of the Twentieth Century

The first half of the twentieth century saw the introduction of the automobile and motorised commercial freight as we know it today.

Vehicles had developed gear systems to manage driving at different speeds and inclines. Synchromesh transmission was invented in 1919 by Earl Avery Thompson and introduced by Cadillac in 1928.

By the outbreak of WW2, traffic management was already organised in most cities. The world's first traffic light was a manually operated gas-lit signal installed in London in December 1868. (It exploded less than a month after it was implemented, injuring its policeman operator). The three-coloured traffic signal appeared first in the city of Detroit in 1920.

Mechanical road traffic control systems were prevalent by the outbreak of WWII, but were very manually controlled mechanisations.

14.6 The Second Half of the Twentieth Century

The self-cancelling indicator system was patented by Werstein in 1951.

While the first automatic transmission using hydraulic fluid was developed in 1932 by two Brazilian engineers, José Braz Araripe and Fernando Lehly Lemos, it was not implemented until they sold the prototype and design to General Motors, who introduced 'hydramatic' transmission. This transmission was introduced in 1939 as options in some Cadillac and Oldsmobile models. The transmission had forward gears and power came from the engine through a fluid coupling. The transmission was not as fast or efficient as a manual but was very popular. By February 1942, when civilian automobile production stopped due to the war, 200 000 had been sold. In 1948, Hydramatic made it to Pontiac, where the take-up rate was 70%. Buick and Chevrolet chose to develop their own automatics, Dynaflow, and Powerglide, respectively. By 1960, 70% of all cars in America had two pedals. The take-up rate in Europe was much slower, due a predeliction in some countries for diesel engines, but much more to the much higher level of taxation on fuel, leading to small engines that adapted less well to the technology, until the latter part of the century, and crowned by the ill-fated British Leyland's invention of the now very successful 'Continuously Variable Transmission' (CVT). In contrast to conventional automatic transmissions, a CVT uses a belt or other torque transmission scheme to allow an 'infinite' number of gear ratios instead of a fixed number of gear ratios, using electrohydraulic means. CVT is now widely used in hybrid vehicles.

In 1965, American Motors (AMC) introduced a low-priced automatic speed control for its large-sized cars with automatic transmissions, and by the 1990 more or less all cars that used automatic transmission also offered cruise control.

By the end of the twentieth century, 'traffic management centres' provided connected means to control traffic in most major cities. At the street level a number of automated, processor controlled, systems replaced treadle, or simple timed, control of traffic lights.

Road tolls commonly used transponders or low-cost bidirectional communications to collect road tolls. The 5.8 GHz 'dedicated short-range communication' was standardised.

14.7   2000–2009

By 2000, serious thought was being given to the concept of the 'connected' vehicle, and, as we have seen from earlier chapters, the architecture for connected cars, and the communications systems and their security, began in earnest.

Electronic road tolling expanded rapidly both as a tool for generating income to pay for development of highway systems, and as an attempt to manage demand.

eCall is a system where an emergency call is generated either automatically via activation of in-vehicle sensors or manually by the vehicle occupants; when activated it provides notification and relevant location information to the most appropriate Public Safety Answering Point, by means of mobile wireless communications networks, and carries a defined standardised minimum set of data notifying that there has been an incident that requires response from the emergency services, and establishes an audio channel between the occupants of the vehicle and the most appropriate Public Safety Answering Point.

eCall was developed in the 2000, and started its path to becoming a standardised and regulated requirement for all new vehicles in 2007. It eventually became a requirement for all new model cars and light vans in 2018, showing how long it can take for innovative new systems to become instantiated.

We saw in Chapter 4, Section 4.2 how C-ITS concepts have been developed throughout the 2000 and continuing up to today (more about that in Section 14.8 below).

The second half of the 2000s saw the introduction of some early 'autonomous' ADAS (advanced driver assistance system) functions, such as rear-view cameras, and adaptive cruise control (ACC), (first fitted in a crude form to some Japanese cars in the 1990, but rolled out quite widely around the world to up-market cars in the latter part of the 2000's.) There was a lot of experimentation with automatic braking, and in 2004 DARPA established the 'Grand Challenge', a competition designed to encourage the development of technologies needed to create the first fully autonomous ground vehicles. The first of these challenges took place in 2004 in Nevada but no team finished the course. The second event was held on October 8, 2005, in southern Nevada and the $2 million prize was won by Stanford University. The third event was held in November 2007, and included a mock urban environment. Driving in traffic and typical vehicle manoeuvres and highway crossings were involved. Carnegie Mellon University in Pittsburgh, claimed the $2 million prize with a converted Chevrolet Tahoe. The race for AVs had begun.

14.8   2010–2019

The past decade has seen much experimentation with AVs, of which, much has been described throughout this book. But for production vehicles, the rollout of 'autonomous' ADAS (that is to say ADAS systems that rely solely on in-vehicle equipment) has been the principal feature in the market place. Electronic stability control (ESC) is now mandatory on all new cars sold in Europe. Lane-keeping systems, ACC, automated emergency braking (AEB), and intelligent speed assistance (ISA) systems are increasingly commonplace, as are automatic headlight dipping, automatic windscreen wipers, traction control, tyre pressure monitoring, etc., now penetrating into average family and compact vehicles.

By the end of the decade, some top-end cars offered some 'automated' features, but still require driver supervision. For example, the Audi A8, has a 'Traffic Jam Pilot' feature capable of navigating highway gridlock at low speeds. Similar systems are known and implemented by others as 'low-speed follow'. The Volvo, BMW, Honda, Mazda, and Cadillac top-end ranges offer a similar system. These systems enable hands-free driving at low speeds on divided highways as long as the driver is paying attention, something the car monitors with a driver-facing camera. The Cadillac system operates at up to highway speeds ('super cruise').

Many systems have been introduced that will keep the car centred in its lane by reading lane markings, by the position of the car ahead (called lane-keeping assist [LKA]; e.g. Acura, Alfa Romeo, Audi, BMW, Buick, Cadillac, Ford, Genesis, Honda, Hyundai, Kia, Infinity, Jaguar, Land Rover, Lexus, Lincoln, Maserati, Mazda, Mercedes, Nissan, Porsche, Subaru, Tesla, Toyota, Volkswagen, Volvo, and others).

Lexus and Toyota 'Lane-Tracing Assist', is similar, but has better reckoning thanks to a higher-performance camera and the ability to see lane markings and the vehicle ahead, as opposed to lane markings only for LKA.

Mercedes says its Active Distance Assist DISTRONIC and Active Steering Assist will be even better at helping the driver keep a safe distance and steer the vehicle. Additionally, speed can also be automatically adjusted in bends and at intersections. Also included are active emergency stop assist and a considerably improved active lane change assist.

Nissan and Infiniti ProPILOT Assist technologies support single-lane driving. Drawing on inputs from radar and camera to read the road ahead and monitor the position of vehicles in front, the system allows the car to react accordingly to assist the driver control acceleration, braking, and steering during single-lane highway driving.

Lane-centring steering goes beyond lane-departure steering assist (which intervenes only as you approach or cross the lane markings) in order to actively centre the vehicle in its lane by tracking lane markings, the vehicle ahead or some combination of the two. Such systems can often negotiate mild curves as well, but nearly all of them require the driver to keep your hands on the wheel, issuing warnings and eventually deactivating if they sense a lack of steering force after a short time.

Hands-free steering centres the car without the driver's hands on the wheel.

Tesla 'Enhanced Autopilot' is in a different league, and pushes 'autonomous' ADAS, has most of the above features. It also offers Autosteer, which assists in steering within a clearly marked lane, and uses traffic-aware cruise control. Under active driver supervision, Tesla FSD – Full Self-Driving Capability – makes steering, acceleration, braking taking into account surrounding traffic. Navigate on Autopilot (offered in beta) actively guides the car from a highway's on-ramp to off-ramp, including suggesting lane changes, navigating interchanges, automatically engaging the turn signal and taking the correct exit; Auto Lane Change Assists in moving to an adjacent lane on the highway when Autosteer is engaged; Autopark helps automatically parallel or perpendicular park your car, with a single touch; summon moves the car in and out of a tight space using the mobile app or key; Smart Summon ensures the car will navigate more complex environments and parking spaces, manoeuvring around objects as necessary to come find you in a parking lot. With Smart Summon, drivers can enable their car to navigate a parking lot and come to them or their destination of choice, as long as their car is within their line of sight.

The Tesla technology takes 'autonomous' ADAS about as far as it can go without becoming AV.

Finally, we have seen many AV demonstrations in the past decade, and throughout this book, I have identified the work still to be done before AV can be implemented on mixed highways. However, by the end of the last decade, low-speed AVs, operating on separated highways, or under very low-speed limitations, have been thoroughly tested and are beginning to be deployed in these limited paradigms.

14.9   2020–2029

The first progress that can be expected is the roll-out of low speed automated shuttles operated largely segregated from free-flowing traffic, and these may be expected to be commonplace by 2025.

The ADAS systems introduced in the previous decade were inherently 'autonomous' relying solely on internal systems. Throughout this book, I have described the commercial and political wrangling regarding the wireless communications for 'connected' vehicles. That argument has slowed the introduction of ADAS systems that involve connectivity with other vehicles and/or the infrastructure. We may expect the communications issue to be resolved one way or another within the next two years. This author's guess is that it is probably likely to prefer G5 5.9 GHz for the introduction of so-called Day 1/1.5 services in 2020, moving to 5G in the latter part of 2020, but it is still a heated debate and might not be the solution adopted. One way or another it is likely to be resolved within the next 24 months.

This means that we can expect the rollout of connected ADAS driver support systems throughout the 2020's. This will introduce systems such as hazardous location notifications; slow or stationary vehicle(s) and traffic ahead warning; road works warning; weather conditions; emergency brake light; emergency vehicle approaching; other hazardous notifications; signage applications, such as vehicle signage and in-vehicle speed limits; signal violation/intersection safety; traffic signal priority request by designated vehicles; green light optimal speed advisory (GLOSA); probe vehicle data; shockwave damping; information on fuelling and charging stations for alternative fuel vehicles; vulnerable road user protection; on-street parking management and information; off-street parking information; park and ride information; connected and cooperative navigation into and out of the city (first and last mile, parking, route advice, coordinated traffic lights); traffic information and smart routing.

So there will be many new features, but they presuppose, or are designed to assist, the driver.

Section 14.4 described the issues delaying the instantiation of AV systems, and concluded that another 5–10 years will be needed before Level 4 and 5 systems can be operational.

So, in practical terms we can expect enhanced connected ADAS systems 2020–2025, and the rollout of increasing Level 1–3 (driver-supervised) AV systems, in the same 2020–2025 period. Finally, the remainder of the technical issues for AV control and safety should be solved in the same 2020–2025 period.

Somewhere between 2025 and 2030 adequate regulatory information will become available. Full Level 4 AVs will become technically feasible from this time, in places where the regulatory information is available – but it will not be universally useable. Places like Sweden, Norway and probably the UK expect to be complete or well advanced nationally, in the case of Sweden and Norway, in respect of major cities in the case of the UK. Other counties will be enabled piecemeal. Significantly, moves by the United States in this area are very unclear, particularly with the administration in power in 2020.

14.10   2030–2039

So if we assume that Level 4 and 5 AVs become widely available circa 2030, what market penetration will they make in the period 2030–2039?

The vehicle parc renews itself over a 20-year period (reflecting an average vehicle life of between 10 and 25 years). Vehicle life has been lengthening as build quality standards have improved. However, the amount of electronics in critical uses in the vehicles and the operational life of electronics will tend to start to reduce vehicle life somewhat. But for the sake of this exercise, let us assume that 5% of the vehicle parc is replaced every year. This means that even if every vehicle sold was an AV, by the end of 2039 only 50% of the vehicle parc would be an AV, and of course, not every vehicle purchase will be an AV. Indeed, as described throughout the book, many people enjoy driving their vehicles. Many more will not trust AVs until they have been on the scene for many years, so a penetration of 30%–50% of new car sales is probably the best that can be expected. This implies that by 2039 probably no more than one-third of the vehicle park will be AVs.

This means that for the whole of the 2030's AVs will be a minority in a mixed parc of vehicles. This has implications in respect of road infrastructure planning and driving behaviour. It also means that the road safety savings – reduced deaths and injuries – will be more slowly realised. Government may therefore be expected to provide incentives for the adoption of AVs. These may take the form of tax breaks, or the infrastructure modified to give advantage to AVs (e.g. reassigning high-occupancy vehicle (HOV) lanes to AV use or otherwise dedicating lanes, even allocating roads to sole AV use).

The complications of the behaviour of human drivers to the presence of AVs will occur (as described throughout the book), and this may further slow AV take-up.

The notion that people will stop wanting to own a vehicle and move to MaaS will be tested. By this time, assuming a vaccine, memories of Covid-19 should have faded and should not be a significant factor.

14.11   2040–2050

Using the basis described above, even by 2050 probably no more than two-thirds of vehicles will be automated, So this decade will also be dominated by a mixed vehicle parc. If that seems overly pessimistic, consider how many 1990 vehicles are still on the roads in 2020, or how, in 1990, you would have envisioned the driving paradigm in 2020. Remember the first picture in this book? We are always hopelessly optimistic about new technology. (The only exception to that being the smartphone). Automatic gear change vehicles became available in the 1950; 70 years later, outside North America, they are still not ubiquitous.

14.12   2050–2060

By this time the comparative safety benefits of AVs should have become clear, and governments may consider regulating against human driven vehicles on safety grounds.

14.13   In Summary

Graphically, we can show this as follows (Figure 14.2):

Whether the projection flattens off in the 2060, or governments consider regulating against human driven vehicles by that stage, is far too speculative to assess at this stage. But the impact of this forecast is that AV vehicles for MaaS will not be available before 2030, and that for the foreseeable future – even to 2060 – will be a slow and steady increase of AVs, but a mixed driving throughout the whole period.

AVs will become steadily more important, but the take-up will be slower than many have hoped.

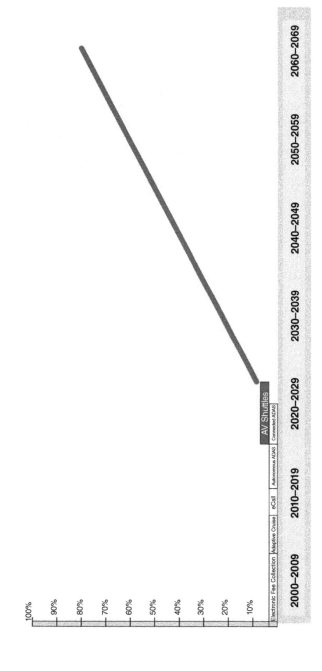

Figure 14.2 Connected and autonomous vehicle (CAV) take-up.

Bibliography

Anderson, J.M., Kalra, N., Stanley, K.D. et al. (2016). *Autonomous Vehicle Technology: A Guide for Policymakers*. Santa Monica, CA: RAND Corporation, www.rand.org/pubs/research_reports/RR443-2.html (accessed 1 November 2016).

Beene, R. (2016). Auto coalition urges U.S. to adapt regulations for autonomous vehicles, Ford, Google, Volvo, Uber and Lyft lead the charge. www.autonews.com/article/20161122/OEM06/161129962/automakers-push-u-sto-adapt-regulations-for-autonomous-vehicles (accessed October 2019).

Blanco, M., Atwood, J., Russell, S. et al. (2016). *Automated Vehicle Crash Rate Comparison Using Naturalistic Data*. Virginia Tech Transportation.

Bou Farah, M., Mercier, D., Delmotte, F., and Lefèvre, E. (2016). Methods using belief functions to manage imperfect information concerning events on the road in VANETs. *Transportation Research Part C: Emerging Technologies* 67 https://www.sciencedirect.com/journal/transportation-research-part-c-emerging-technologies.

Brussels, 30.11.2016 COM (2016). 766 Final communication from the commission to the European parliament, the council, the European economic and social committee and the committee of the regions. A European strategy on cooperative intelligent transport systems, a milestone towards cooperative, connected and automated mobility, https://ec.europa.eu/energy/sites/ener/files/documents/1_en_act_part1_v5.pdf (accessed October 2019).

Campolo, C., Molinaro, A., and Scopigno, R. (2015). From today's VANETs to tomorrow's planning and the bets for the day after. *Vehicular Communications* 2 (3): 158–171.

Casner, S.M., Geven, R.W., and Williams, K.T. (2013). The effectiveness of airline pilot training for abnormal events. *Human Factors* 53 (3): 477–485. www.ncbi.nlm.nih.gov/pmc/articles/PMC6109944/.

Casner, S.M., Geven, R.W., Recker, M.P., and Schooler, J.W. (2014). The retention of manual flying skills in the automated cockpit. *Human Factors* 56 (8): 1506–1516. https://journals.sagepub.com/doi/10.1177/0018720814535628.

Casner, S.M., Hutchins, E.L., and Norman, D. (2016). The challenges of partially automated driving. *Communications of the ACM* 59 (5) http://delivery.acm.org/10.1145/2840000/2830565/p70-casner.pdf?ip=86.145.149.51&id=2830565&acc=OA&key=4D4702B0C3E38B35%2E4D4702B0C3E38B35%2E4D4702B0C3E38B35%2E47248115B7DBE78F&__acm__=1573375490_1491281e56995c1ab9c8c1c790850b65.

Automated Vehicles and MaaS: Removing the Barriers, First Edition. Bob Williams.
© 2021 John Wiley & Sons Ltd. Published 2021 by John Wiley & Sons Ltd.

C-ITS Platform (2016). C-ITS Platform Final report. http://ec.europa.eu/transport/sites/transport/files/themes/its/doc/c-its-platform-finalreport-january-2016.pdf (accessed October 2019).

COM (2015) 192 Final communication from the commission to the European parliament, the council, the European economic and social committee and the committee of the regions. A digital single market strategy for Europe. https://eur-lex.europa.eu/legal-content/EN/TXT/PDF/?uri=CELEX:52015DC0192&from=EN (accessed October 2019).

Commercial Fleet News (2019). www.commercialfleet.org/news/truck-news/2019/02/04/mercedes-switches-focus-away-from-platooning-trials (accessed October 2019).

Cunha, F., Villas, L., Boukerche, A. et al. (2016). Data communication in VANETs: protocols applications and challenges. *Ad Hoc Networks* 44: 90–103.

Curlander, J.C., Russell, R.S. and Bathurst, A.S. et al. (2017). Lane assignments for autonomous vehicles. US Patent US 9,547,986 B1. https://patents.google.com/patent/US9547986B1/en (accessed September 2020).

Dungs, J., Herrmann, F., and Duwe, D. et al. (2016). The value of time, Potential for user-centered services offered by autonomous driving, Stuttgart. www.iao.fraunhofer.de/lang-en/index.php?option=com_content&view=article&id=1266&back=1203 name (accessed September 2020).

EC Joint Research Council (2018). Certificate Policy For Deployment and Operation of European Cooperative Intelligent Transport Systems (C-ITS). https://ec.europa.eu/transport/sites/transport/files/c-its_certificate_policy-v1.1.pdf (accessed October 2019).

ETSC (2016). Briefing "Prioritising the safety potential of automated driving in Europe." https://etsc.eu/wp-content/uploads/2016_automated_driving_briefing_final.pdf (accessed September 2020).

Europe Autonews.com (2019). https://europe.autonews.com/article/20180808/ANE/180809840/france-pushes-for-highly-automated-vehicles-by-2022 (accessed October 2019).

European Commission (2014). Special Eurobarometer 422a, Quality of transport. https://ec.europa.eu/commfrontoffice/publicopinion/archives/ebs/ebs_422a_en.pdf (accessed September 2020).

European Commission (2015a). G7 declaration on automated and connected driving. http://ec.europa.eu/commission/20142019/bulc/announcements/g7-declaration-automated-and-connected-driving_en (accessed October 2019).

European Commission (2015b). Network and information security directive: co-legislators agree on the first eu-wide legislation on cybersecurity. https://ec.europa.eu/digital-single-market/en/news/network-andinformation-security-directive-co-legislators-agree-first-eu-wide-legislation (accessed October 2019).

European Commission (2015c). Harmonized security policies for cooperative Intelligent Transport Systems create international benefits. https://ec.europa.eu/digital-single-market/en/news/harmonized-securitypolicies-cooperative-intelligent-transport-systems-create-international (accessed October 2019).

European Commission (2016a). Strategy for Low-Emission Mobility https://ec.europa.eu/transport/themes/strategies/news/2016-07-20-decarbonisation_en (accessed October 2019).

European Commission (2016b). Commission launches GEAR 2030 to boost competitiveness and growth in the automotive sector. http://ec.europa.eu/growth/

toolsdatabases/newsroom/cf/itemdetail.cfm?item_id=8640&lang=en (accessed October 2019).

European Commission (2017). DG move smart mobility and services https://ec.europa.eu/transparency/regexpert/index.cfm?do=groupDetail.groupDetailDoc&id=34596&no=1 (accessed October 2019).

European Commission, Joint Research Centre (2017). The r-evolution of driving: from connected vehicles to coordinated automated road transport (C-ART): part I: framework for a safe & efficient coordinated automated road transport (C-ART) system.

European Union (2016). The declaration of amsterdam, cooperation in the field of connected and automated driving. https://english.eu2016.nl/documents/publications/2016/04/14/declaration-ofamsterdam (accessed 13 March 2019).

Evensen, K. (2020). ISO TC204 WG19. Unpublished paper.

Fagnant, D.J. and Kockelman, K.M., (2014). The travel and environmental implications of shared autonomous vehicles, using agent-based model scenarios. *Transportation Research Part C*, 40: 1–13.

French manufacturer Renault (2019). https://life.renault.co.uk/innovation/autonomous-vehicles (accessed October 2019).

HARTS ITS Architecture (2019). http://htg7.org/index.html (accessed October 2019).

Kesting, A. and Treiber, M. (2008). How reaction time, update time, and adaptation time influence the stability of traffic flow. *Computer-Aided Civil Infrastructure Engineering* 23 https://doi.org/10.1111/j.1467-8667.2007.00529.x.

Kröger, F. (2016). *Automated driving in its social, historical and cultural contexts*. In: *Autonomous Driving in Its Social, Historical and Cultural Contexts*, 41–68. Cham: Springer.

Kühn, M. (2016). Takeover times in highly automated driving, Gesamtverband der Deutschen Versicherungswirtschaft e.V. (GDV). http://indexsmart.mirasmart.com/25esv/PDFfiles/25ESV-000027.pdf (accessed October 2019).

LaMondia, J.J., Fagnant, D.J., Qu, H. et al. (2016). Long-distance travel mode shifts due to automated vehicles: a statewide mode-shift simulation experiment and travel survey analysis. TRB Paper No. 16-3905, 95th Annual Meeting of the Transportation Research Board. Washington, DC.

Langenwalter, J. (2016). Artificial intelligence in autonomous driving. www.electronics-eetimes.com/designcenter/artificial-intelligence-autonomous-driving-0 (accessed October 2019).

Lawson, P. et al. (2015). Creative commons attribution non commercial 2.5 Canada license, CC BY-NC 2.5 CA. https://creativecommons.org/licenses/by-nc/2.5/ca/ (accessed October 2019).

Le Vine, S., Zolfaghari, A., and Polak, J. (2015). Autonomous cars: the tension between occupant-experience and intersection capacity. *Transportation Research Part C: Emerging Technologies* www.journals.elsevier.com/transportation-research-part-c-emerging-technologies.

Litman, T. (2017) Autonomous vehicle implementation predictions implications for transport planning

McCarthy, C., Harnett, K., Carter, A., and Hatipoglu, C. (2014). Assessment of the information sharing and analysis center model, Report No. DOT HS 812 076. Washington, DC: National Highway Traffic Safety Administration.

McKinsey (2016) Monetizing car data: new service business opportunities to create new customer benefits. www.mckinsey.com/~/media/McKinsey/Industries/Automotive %20and%20Assembly/Our%20Insights/Monetizing%20car%20data/Monetizing-car-data.ashx (accessed September 2020).

Millard-Ball, A. (2016). Pedestrians, autonomous vehicles, and cities. *Journal of Planning Education and Research*: 1–7 https://doi.org/10.1177/0739456X16675674. http://bit.ly/2ga4vV0.

Ngoduy, D. (2013a). Analytical studies on the instabilities of heterogeneous intelligent traffic flow. *Communications in Nonlinear Science and Numerical Simulation* 18 https://doi.org/10.1080/21680566.2014.960503.

Ngoduy, D. (2013b). Instabilities of cooperative adaptive cruise control traffic flow: a macroscopic approach. *Communications in Nonlinear Science and Numerical Simulation* 18 www.tandfonline.com/ https://doi.org/10.1080/21680566.2016.1142401.

NHTSA (2016). Federal automated vehicles policy accelerating the next revolution in roadway safety. https://one.nhtsa.gov/nhtsa/av/pdf/Federal_Automated_Vehicles_Policy.pdf (accessed October 2019).

Nowacki, G. (2011). Development and standardization of intelligent transport systems. In: Transport Systems and Processes, Marine Navigation and Safety of Sea Transportation (ed. T. Neumann), 156–167. CRC Press.

OECD/ITF (2015). Automated and autonomous driving, regulation under uncertainty. www.itfoecd.org/sites/default/files/docs/15cpb_autonomousdriving.pdf (accessed October 2019).

Okumura, B., James, M.R., and Kanzawa, Y. et al. (2016). Challenges in perception and decision making for intelligent automotive vehicles: a case study. IEEE Transactions on Intelligent Vehicles. https://ieeexplore.ieee.org/document/7448943 (accessed September 2020).

Sabur, R. (2017). Driverless cars will cause congestion on Britain's roads to worsen for years, study finds. www.telegraph.co.uk/news/2017/01/06/driverless-cars-will-cause-congestionbritains-roads-worsen (accessed June 2017).

San Francisco County Transportation Authority (2018). TNCs & Congestion. www.sfcta.org/emerging-mobility/tncs-and-congestion (accessed October 2019).

Sarter, N.B., Mumaw, R.J., and Wickens, C.D. (2007). Pilots' monitoring strategies and performance on automated flight decks: an empirical study combining behavioural and eye-tracking data. *Human Factors* 49 (3): 347–357. www.ncbi.nlm.nih.gov/pubmed/17552302.

Schoettle, B. and Sivak, M. (2014). A survey of public opinion about autonomous and self driving vehicles in the U.S., the U.K., and Australia. Technical report.

Schoettle, B. and Sivak, M. (2015). Should we require licensing tests and graduated licensing for self-driving vehicles?, Report UMTRI-2015-33, Transportation Research Institute, University of Michigan. www.umich.edu/~umtriswt (accessed October 2019).

Seba, T. (2016). *Clean Disruption - why Energy & Transportation Will Be Obsolete by 2030*. Oslo, Norway: Swedbank Nordic Energy Summit.

Sivak, M. and Schoettle, B., (2015). Potential impact of self-driving vehicles on household vehicle demand and usage, UMTRI-2015-3. www.umtri.umich.edu/our-results/publications/potential-impact-self-driving-vehicles-household-vehicle-demand-and-usage (accessed September 2020).

Smith, B.W. (2013). Human error as a cause in vehicle crashes. http://cyberlaw.stanford
.edu/blog/2013/12/human-error-cause-vehicle-crashes (accessed October 2019).

Smith, B.W. (2014). *A Legal Perspective on Three Misconceptions in Vehicle Automation*
Road Vehicle Automation. New York: Springer.

Society of Automotive Engineers (2017). J 3016, taxonomy and definitions for terms related
to on-road motor vehicle automated driving systems.
www.sae.org/standards/content/J3016 201806 (accessed 10 April 2017).

Sun, Y., Wu, L., Wu, S. et al. (2015). Security and privacy in the internet of vehicles. https://
ieeexplore.ieee.org/document/7428337 (accessed September 2020).

Tesla (2019). www.tesla.com/en_GB/autopilot (accessed October 2019).

The Daimler/Mercedes website (2019). www.daimler.com/innovation/autonomous-driving
(accessed October 2019).

The Ford Motor company website (2019). https://corporate.ford.com/articles/products/
autonomous-2021.html (accessed October 2019).

The Guardian (2019). www.theguardian.com/world/2017/nov/23/philip-hammond-
pledges-driverless-cars-by-2021-and-warns-people-to-retrain (accessed October 2019).

The Independent (2019). www.independent.co.uk/news/uk/home-news/driverless-cars-
uk-roads-2019-self-driving-hacking-cyber-security-a8766716.html (accessed October
2019).

The Verge (2019). www.theverge.com/autonomous-cars (accessed October 2019).

Tillemann, L. and McCormick, C. (2016). Will driverless-car makers learn to share? www
.newyorker.com/business/currency/will-driverless-car-makers-learn-to-share (accessed
October 2019).

Townsend, E. (2016). *Prioritising the Safety Potential of Automated Driving in Europe.*
European Transport Safety Council. http://etsc.eu/wpcontent/uploads/2016_
automated_driving_briefing_final.pdf (accessed 10 April 2017).

UK Department for Transport: The future of mobility (2019). https://assets.publishing
.service.gov.uk/government/uploads/system/uploads/attachment_data/file/786654/
future-of-mobility-strategy.pdf (accessed October 2019).

UK Parliament (2019). Mobility as a service. Eighth Report of Session 2017-19 https://
publications.parliament.uk/pa/cm201719/cmselect/cmtrans/590/full-report.html
(accessed October 2019).

UN ECE Transport: Sustainable Urban Mobility and Public Transport (2015). www.unece
.org/fileadmin/DAM/trans/main/wp5/publications/Sustainable_Urban_Mobility_and_
Public_Transport_FINAL.pdf (accessed October 2019).

UNECE (2019). WP29 GVRA Framework document on automated/autonomous vehicles
ECE/TRANS/WP.29/2019/34/Rev.1 "revised framework document on
automated/autonomous vehicles" www.unece.org/fileadmin/DAM/trans/doc/2019/
wp29/WP29-177-19e.pdf (accessed October 2019).

Volvo website (2019). www.volvocars.com/en-kw/own/own-and-enjoy/autonomous-
driving (accessed October 2019).

Weber, M. (2014). Where to? A history of autonomous vehicles. www.computerhistory
.org/atchm/where-to-a-history-ofautonomous-vehicles (accessed October 2019).

WHO (2015). Global status report on road safety 2015. www.who.int/violence_injury_
prevention/road_safety_status/2015/status_report2015/en/ (accessed September 2020).

WHO (ed.) (2018). Global status report on road safety 2018 (pdf) (official report). Geneva:
World Health Organisation (WHO). Tables A2 & A11.

Wiener, E.L. (1985). Human factors of cockpit automation: a field study of flight crew transition. NASA Contractor Report 177333. NASA, Moffett Field, CA, http://ntrs.nasa .gov/archive/nasa/casi.ntrs.nasa.gov/19850021625.pdf (accessed October 2019). (accessed October 2019)

SMMT (2019). The UK society of motor manufacturers and traders (SMMT) issued a report. Connected and autonomous vehicles 2019 report/winning the global race to market. www.smmt.co.uk/wp-content/uploads/sites/2/SMMT-CONNECTED-REPORT-2019.pdf (accessed September 2020).

GEAR 2030, 2016, GEAR 2030 Discussion paper WG1 Adaptation of the EU value chain. https://circabc.europa.eu/sd/a/795a8571-63a8-4077-8315-a944b6d88787/Discussion %20Paper%20%20Adaptation%20of%20the%20EU%20value%20chain%2008-01-2016 .pdf (accessed October 2019).

GEAR 2030, 2016, GEAR 2030 Discussion Paper WG2, Roadmap on Highly Automated vehicles. https://circabc.europa.eu/sd/a/a68ddba0-996e-4795-b207-8da58b4ca83e/ Discussion%20Paper%C2%A0%20Roadmap%20on%20Highly%20Automated %20Vehicles%2008-01-2016.pdf (accessed October 2019).

Business Insider 2019/10. www.businessinsider.com/volkswagen-adds-vehicle-to-vehicle-communication-to-golf-lineup-2019-10?international=true&r=US&IR=T (accessed October 2019).

Autonews (2019). www.autonews.com/cars-concepts/vw-touts-connectivity-leap-new-golf?utm_source=Triggermail&utm_medium=email&utm_campaign=Post%20Blast %20bii-transportation-and-logistics:%20VW%20adds%20V2V%20communication%20to %20upcoming%20Golf%20lineup%20%7C%20Shipwell%20raises%20%2435M%20to%20 expand%20its%20reach%20%7C%20XPO%20Logistics%20partners%20with%20MIT &utm_term=BII%20List%20T%26L%20ALL (accessed October 2019).

ISO 21177, Intelligent transport systems — ITS-station security services for secure session establishment and authentication between trusted devices. www.iso.org/standard/70056 .html (accessed September 2020).

Wired. 2015/07. www.wired.com/2015/07/hackers-remotely-kill-jeep-highway/ (accessed October 2019).

UNECE. October 2019. Collaboration and common approaches between WP.1-WP.29 on automated vehicles. WP.29–179-05. www.unece.org/fileadmin/DAM/trans/main/wp29/ WP29-179-05e.pdf (accessed October 2019).

UK: Automated and Electric Vehicles Act, 2018. www.legislation.gov.uk/ukpga/2018/18/ contents/enacted (accessed October 2019).

European Parliament. Directive 2006/126/EC of the European parliament and of the council of 20 December 2006 on driving licences. https://eur-lex.europa.eu/LexUriServ/ LexUriServ.do?uri=OJ:L:2006:403:0018:0060:EN:PDF (accessed October 2019).

European Parliament. Directive 2009/103/EC of the European parliament and of the council of 16 September 2009 relating to insurance against civil liability in respect of the use of motor vehicles, and the enforcement of the obligation to insure against such liability. https://eur-lex.europa.eu/legal-content/EN/TXT/PDF/?uri=CELEX: 32009L0103&from=EN (accessed October 2019).

eenews Automotive. (2019) www.eenewsautomotive.com/content/electronic-control-system-partitioning-autonomous-vehicle (accessed October 2019).

European Council. (85/374/EEC) Council Directive of 25 July 1985 on the approximation of the laws, regulations and administrative provisions of the Member States concerning liability for defective products. https://eur-lex.europa.eu/legal-content/EN/TXT/PDF/?uri=CELEX:31985L0374&from=EN (accessed October 2019).

UK Government. 1945. UK: Law Reform (Contributory Negligence) Act 1945 Law Reform (Contributory Negligence) Act 1945. www.legislation.gov.uk/ukpga/Geo6/8-9/28/data .pdf (accessed Ocrober 2019).

GEAR 2030, 2016, GEAR 2030 discussion paper WG2, roadmap on highly automated vehicles. https://circabc.europa.eu/sd/a/a68ddba0-996e-4795-b207-8da58b4ca83e/Discussion%20Paper%C2%A0-%20Roadmap%20on%20Highly%20Automated %20Vehicles%2008-01-2016.pdf (accessed October 2019).

US ITS Architecture ARC-IT version 8.3. https://local.iteris.com/arc-it/ (accessed October 2019).

TRAMAN (2019). TRAMAN21 (Traffic Management for the 21st Century) www .traman21.tuc.gr/ (accessed October 2019).

WHO (2010). World report on road traffic injury prevention. World Health Organisation.

European Parliament. Directive 2010/40/EU of the European parliament and of the council of 7 July 2010 on the framework for the deployment of Intelligent Transport Systems in the field of road transport and for interfaces with other modes of transport.

FORD (2019). https://media.ford.com/content/dam/fordmedia/pdf/Ford_AV_LLC_ FINAL_HR_2.pdf.

Transport for London (TfL) performance report third quarter of 2015/16. http://content.tfl .gov.uk/q3-16-17-quarterly-performance-report.pdf (accessed September 2020).

European Parliament. Regulation (EU)2016/679 of the European parliament and of the council of 27 April 2016 on the protection of natural persons with regard to the processing of personal data and on the free movement of such data, and repealing Directive 95/46/EC (General Data Protection Regulation). https://eur-lex.europa.eu/eli/ reg/2016/679/oj (accessed September 2020).

European Commission (2019/October). C-ITS deployment takes off, increasing road safety and decreasing congestion. C-Roads. www.c-roads.eu/fileadmin/user_upload/media/ Dokumente/Statement_Continued_C-ITS_deployment_in_Europe_based_on_available_ and_proven_technology_V1.2.pdf (accessed October 2019).

NTSB (2018). Preliminary report – highway – HWY18MH010 (pdf) (The information in this report is preliminary). National Transportation Safety Board.

NTSB (2018). NTSB update: uber crash investigation (Press release). National Transportation Safety Board.

Wikipedia (2019). https://en.wikipedia.org/wiki/Death_of_Elaine_Herzberg (accessed October 2019).

BSB (2016). Legal consequences of an increase in vehicle automation, Part 1. Bundesanstalt für Straßenwesen Brüderstraße 53, D-51427 Bergisch Gladbach.

Yap, M.D., Correia, G., and van Arem, B. (2016). Preferences of travellers for using automated vehicles as last mile public transport of multimodal train trips. *Transportation Research Part A: Policy and Practice* 94 https://daneshyari.com/article/ preview/4929013.pdf.

Index

Automated Vehicles and MaaS: Removing the Barriers, First Edition. Bob Williams.
© 2021 John Wiley & Sons Ltd. Published 2021 by John Wiley & Sons Ltd.